UPRIGHT

Books by Craig Stanford

Significant Others

Meat-Eating and Human Evolution

The Hunting Apes

Chimpanzee and Red Colobus

UPRIGHT

The Evolutionary Key to Becoming Human

CRAIG STANFORD

Houghton Mifflin Company
BOSTON
NEW YORK
2003

Visit our Web site: www.houghtonmifflinbooks.com.

Library of Congress Cataloging-in-Publication Data

Stanford, Craig B. (Craig Britton), date.
Upright : the evolutionary key to becoming human / Craig Stanford.
p. cm.
Includes bibliographical references and index.
ISBN 0-618-30247-6
1. Human evolution. 2. Bipedalism. I. Title.

GN282.5.S73 2003
599.93'8 — dc21 2003050858

Printed in the United States of America

MP 10 9 8 7 6 5 4 3 2 1

Book design by Anne Chalmers
Typefaces: Janson, Trade Gothic Condensed

To Jane and Paul Moore

I have met with but one or two persons in the course of my life who understood the art of Walking.

— Henry David Thoreau, *Walking*

CONTENTS

ACKNOWLEDGMENTS

Writing a book that attempts to interpret the work of my colleagues is a daunting challenge, because it is bound to receive an unfavorable reaction from at least half the people whose opinions I value highly. This is especially true when writing about the human fossil record. The sample of specimens is small, the scientific opinions deeply held, and the debates fierce. I sent parts of the manuscript for *Upright* to my colleagues with more trepidation than anything I have done since sitting down for my graduate school oral examinations many years ago. But my fears were groundless. The comments that I received from the major combatants in the field were overwhelmingly intended to help the book, not hurt it, and the constructive criticism that I thereby obtained was the key to finishing it.

For their responses to my requests for advice on the manuscript and its interpretations, I am grateful to Tim White of the University of California, Berkeley, who answered my lengthy e-mailed queries within hours with detailed deconstruction of some of my more mistaken facts and reasoning. C. Owen Lovejoy of Kent State University promptly sent unpublished papers and a CD filled with information that I had missed. Russell Tuttle of the University of Chicago did the same and once again helped me edit an unruly manuscript. William McGrew of Miami University, for years the best constructive critic a scientist could hope for, offered many cautionary notes to my broad conclusions. Misia Landau of Harvard Medical School, who sparked the ideas for some of the earliest parts of the book in a series of conversations several years ago, pointed me to the writings about bipedalism by historical figures as unlikely as Friedrich Engels. And my long-time colleague, coauthor, and friend, John Allen of the University of Iowa, provided his unique perspective on the whole thing, as usual. I had other helpful discussions about upright walking with Kevin Hunt of Indiana University and Monte McCrossin of New Mexico State University. I especially thank my editor at Houghton Mifflin, Laura van Dam, and my agent, Russell Galen, of Scovil, Chichak and Galen Literary Agency, Inc.

My initial interest in the origins of bipedalism was sparked by the chimpanzees of my study group in Bwindi Impenetrable National Park, Uganda. They live in a threatened habitat more beautiful than a scientist could hope to spend time in. One day those chimps stood upright in a large tree and began to eat the figs hanging overhead as casually as you might reach for the top shelf in the supermarket. They have since proved to be more bipedal than any apes that I have watched anywhere else, walking upright often but always in the treetops. Watching this made it clear to me that some time-honored theories to explain standing upright were all wrong. I sat down one morning to write a scholarly review paper outlining current thinking about great ape models of the rise of human walking. As the paper grew toward book length several months later, I realized that *Upright* would be the result.

I am grateful for the generous funding that has allowed me to continue my research at Bwindi, now in its seventh year, especially from the National Geographic Society, the Wenner Gren Foundation for Anthropological Research, the Fulbright Foundation, the L.S.B. Leakey Foundation, and the Jane Goodall Research Center of the University of Southern California. The government of Uganda (specifically, the Ugandan Wildlife Authority and the Uganda National Council for Science and Technology) has provided permission to work in that country.

I'm also grateful for the support of many people in Africa and at home who helped with the project, the ensuing publications, and this book, especially, John Bosco Nkurunungi, Caleb Mgamboneza, Mitchell Keiver, Dr. Alastair McNeilage, and Gervase Tumwebaze in Uganda; Jane Goodall in Tanzania; and Christopher Boehm, Adriana Hernandez, Erin Moore, Terrelita Price, Tatiana White, Adam Stanford-Moore, Gaelen Stanford-Moore, and Marika Stanford-Moore in California.

Errors that remain in the book despite all the help offered by my friends, colleagues, and family are, of course, my own.

PREFACE
BABY STEPS

I REMEMBER VIVIDLY the first time that each of my three children took her or his first unassisted steps. My firstborn had been "cruising" for weeks — pulling herself up and walking while holding on to furniture, people, dogs, and anything else that she could grab. But at ten months she was ready to be a biped. She stepped away from my hands and walked several lock-legged goose steps into her mother's arms. My daughter's wide eyes showed her shock at the performance. We beamed, imagining that our parenting skills had something to do with teaching her this most natural of all uniquely human acts. Three years later we were living in a village in rural Mexico and obsessing about the diseases that our younger daughter was contracting by crawling in the dust. Then one day she stood up and toddled, and that was that. My son was a different story; I was in East Africa, having left on a monthlong trip knowing I would likely miss the big event. Sure enough, shortly after I arrived in Uganda, I learned through the crackling static of a phone call that Adam, after much frustration at trying to carry a ball while crawling, had simply stood up and walked, the ball in his arms and an ear-to-ear grin of accomplishment on his face.

Few of us appreciate our history of becoming bipeds, perhaps because walking requires so little energy or thought. Most of us think that our exalted intellect or our ability to grasp with our thumbs is what sets us apart from the other primates. But all primates share the grasping thumb, and the difference between an ape's brain and our own is not as great as people think. Some parts have undergone a critical reorganization, such as the speech centers, but a human brain is basically a ballooned version of a chimpanzee brain.

Our ability to stand and walk habitually on two feet, however, represents a fundamental change from the *kind* of creatures that our ancestors were. Bipedalism preceded the expansion of brain size by about five million years; it truly announced the dawn of humanity. Becoming bipedal made us human. Whenever a fossil human is discovered, the first piece of crucial information that everyone wants to know is "Did it walk upright?" The second question is "How will it change our family tree?"

To an extent unappreciated by most of us, walking is sexy. It is the key part of a cascade of traits that evolved together in an intricate mosaic of ape and early human features. For instance, walking on two legs rather than four released our bodies from the constraints of the synchronized breathing gait that so many other animals, such as dogs and horses, live by. Once the lungs of our two-legged ancestors were freed, they could modulate their breathing in subtle ways that may have contributed to the evolution of speech. The connection between walking upright and speaking is one of many vivid examples of the jigsaw-puzzle evolution of our bodies.

Why we are bipedal is not simple to explain. In this book I show that the question really consists of two parts: what made our ancestors take their first steps, and what the evolutionary impetus was for those "toddlers" to become highly efficient marathon walkers and runners. The first steps were, according to the latest research, mere shuffles that helped our simian ancestors reach plant foods such as figs that were just beyond their grasp. Then these marginal bipeds found a light at the end of the endless search-for-energy tunnel: meat. The meat came in the form of both small game animals, which could be captured and eaten raw, and the carcasses of animals, both small and large, for which the Promethean humans fanned out in search of each day.

Meat eating provided a new and crucial source of protein, fat, and calories that may have enabled the evolution of bigger brains and helped our cognitive abilities to evolve. As meat eating became more important, our ancestors adopted new ways of life that resulted in a hominid that began to rule the planet. Being bipedal did not, contrary to popular conceptions, lead directly to brain expansion; the two events occurred millions of years apart in evolutionary time.

The traditional view of human origins goes something like this: Six million years ago an ape ancestor left the comfort and security of the African forests for life on the savanna. Its new home offered many opportunities for advancement — including open country and a meat-rich diet that the old home lacked. The ape evolved a means of travel in which standing upright became not just a quick periscope but a way of life. Upright posture allowed our ancestors to carry tools, chunks of butchered carcass, and even babies. But the trade-off was enormous. Predators of every shape and size, from leopards to saber-toothed cats, wandered the grass searching for prey all day and night. The new stance left the emerging human without a means of rapidly escaping predators. The single advantage that allowed early humanity to survive, and turned the tide in favor of our lineage, was a rapidly expanding brain. Armed only with its wits, the runty little human eked out an existence for millions of years, eventually prospering and sending its progeny into the present as big-brained *Homo sapiens.*

The familiar image of our ancestors' progression through stages of hunched-over, shuffling, apelike creatures into humans is an appealing one, but each of its elements is being called into question. The idea that we slowly evolved toward perfection is as wrong as it is entrenched. Animals don't evolve *toward* anything; natural selection molds them generation by generation. At each stage the animal's form must be efficiently designed so that it may succeed at eating, rearing offspring, and so on, or natural selection eliminates that animal's genes from the next generation. We are not at the apex of life's intricate evolution from our ape forebears, no matter what both popular and scientific accounts claim.

BIPEDS ARE BIZARRE

Standing on two feet is a bizarre posture and an even more bizarre way to walk. Of the more than two hundred species of primates on earth today, one is bipedal. Of more than 4,000 species of mammals, one — the same one — is fully bipedal when walking (a few oddities such as kangaroo rats and meerkats stand bipedally for a few moments at a time). If we include thousands more kinds of animals — such as amphibians and reptiles — walking on two feet emerges as the most unlikely way to get around. Kangaroos and birds such as ostriches and penguins are bipedal — sort of. But they are built on an entirely different body plan and are not, strictly speaking, reliant only on their legs for transport. Even if we throw in all the extinct forms of terrestrial animal life, such as *Tyrannosaurus rex* and its kin, the percentage of bipeds is still remarkably small. And birds and dinosaurs differ markedly in their brand of upright posture. Most birds have stiff and relatively short tails. They maintain stability by having their center of mass far forward from the pelvis; this forward gravity center necessitates standing with the upper leg bone bent. Birds that have adopted flightlessness, like ostriches, generate their power stride by rotating the lower leg around the knee joint. Upright dinosaurs like *Allosaurus* or *Velociraptor* opted for a center of gravity near the pelvis and rotated their entire leg during striding.

The reason that upright posture and walking arose is the most fundamental question in human evolution. It begets critical puzzles, such as why bipedalism hasn't evolved many more times, and whether the evolution of our unique posture and gait is connected to our massive brain and extraordinary intelligence.

. . .

In this book I emphasize that humankind is only a twig on an evolutionary bush rather than the top rung of an evolutionary ladder of excellence. The fossil record for the rise of bipedalism has just begun to tell us that even as protohumans diverged from the apes, bipedalism existed in a variety of forms. In 2000, for instance, researchers in Kenya announced the discovery of an early human fossil that they named *Kenyanthropus platyops*, which appears to have been a contemporary of other early human species, such as that to which the famous fossil human commonly known as Lucy belongs. Until that discovery, we believed that our family tree had only one trunk. An even more recent and controversial find, "Toumai," is a primitive fossil from the Sahara Desert that some experts believe represents the earliest known member of the human family.

These are heady times in fossil hunting. We are learning that a wide variety of evolutionary experiments, using bipedalism as a recurring theme, took off about five million years ago. Most failed. One branching lineage survived to the present. Also, our early ancestors were not poor bipeds who evolved slowly into "good" two-legged walkers. Emerging evidence suggests that a menagerie of species existed with a variety of characteristics, and they did not form one linear progression from "primitive" to "advanced" bipeds. Our obsession with linear progress has led us severely astray in solving the riddle of why we became bipedal.

How we walk today comes from a cornucopia of evolutionary forces at work on our ancestors' bodies. The modern architecture of the spine, pelvis, feet, and hands, and even nervous and circulatory systems, follows directly from the conversion from quadrupedalism to bipedalism. Other changes, not preserved in stone but equally important in our ability to stand and walk on two legs, took place in our behavior. Our apelike ancestors lived in the forest, climbed in trees, and ate fruits and leaves and occasionally meat. As the hominid emerged from the forest, it underwent changes in foraging strategies, diet, preferred habitat, and tool technologies. The hominid's mating system and social life are unknown to us, although we can make some reasonable inferences. And from this ape ancestor came one with a cerebral volume that was only marginally larger but had an entirely new way of walking. No doubt, changes in social behavior contributed to cognitive changes. Tool technologies changed too, expanding the resources available to this population. Because all these new aspects of emerging humanity cascaded one upon the other, teasing out which caused which is difficult. The intricate puzzle of our humanity came into focus as new pieces were added to old ones, slowly changing our ancestors from one thing to another.

How we became bipedal is a chronicle of how we became human. As our way of moving about changed, so did our niche in the world, our perspective, and our prospects. This chronicle is also an argument for why we must move our view of the earliest stages of humanity from old-fashioned notions of progress and linearity into a more modern Darwinian sensibility. The debates in human evolution research are fierce, because the fossils are few and far between, and because their implications are far-reaching. I will try to convey a sense of the science and the scientific politics that drive the process of making and breaking theories, as well as recent research that has uncovered a variety of key pieces of human ancestry. The story at the heart of this book is truly an odyssey, made more fantastic because it actually occurred.

1 A FIRST STEP

RAYMOND DART, professor of anatomy at the University of the Witswatersrand in South Africa, faced a dilemma that day in 1924. Two burly men wearing the uniforms of South African Railways were coming up the driveway of his Johannesburg house. They were lugging two large wooden crates, cherished prizes that Dart had been anxiously awaiting for the past week. But the timing could not have been worse: Dart was watching the delivery from the window of his bedroom, where he was struggling into an expensive morning coat. A colleague was about to be married in Dart's parlor, and the professor was best man.

Should he open the boxes or finish dressing? He ripped off his stiff collar and rushed to the door to take delivery of the crates, which had been sent by a Mr. Spiers, the manager of an up-country limestone quarry at a place called Taung. Dart, an avid fossil collector, had learned from a geologist that Spiers's commercial limestone quarry had turned up some interesting fossils. The manager had complied with the geologist's request to ship the specimens to Dart.

Ignoring his wife's protests that his colleague's wedding should take precedence, Dart grabbed a crowbar from the garage and pried the lid off the first box. He was disappointed: The crate held scattered fragments of fossilized turtle shells, turtle eggs, and unidentifiable fossilized bones. Then he removed the lid of the second box. The geologist had promised material of interest, and just the previous week he had shown Dart the fossilized skull of an extinct baboon, the first such primate ever recorded in sub-Saharan Africa. Surely this box would contain better fossils than turtle eggs. The dusty objects in that second crate were the most important piece of evidence of human origins to date. The skull and mandible of a child were nestled there, and the skull's anatomy was clearly not that of a mere monkey. After the wedding Dart set to work, separating the skull from the rock matrix in which it sat. When he freed it, he knew at once that what he beheld was a new species of human, not an ape.

The beauty of the Taung finding lay not just in its importance but in the specimen itself. The interior of the skull had fossilized along with the face, so a perfect mold of the brain lay inside the open brain case. In the next few days Dart and a colleague painstakingly removed rock matrix from the skull, revealing a childlike face and a mouth studded with tiny peglike milk teeth. This characteristic of the Taung child, as the specimen came to be nicknamed, stood out for Dart, because the teeth were nothing like those of a baboon or chimpanzee. They were fairly uniform in size and clearly those of a youngster. And the face was like ours; it lacked both robust bony ridges above the eyes and a protruding apelike snout.

Within a few weeks Dart mailed a manuscript to the prestigious science journal *Nature*. He had christened the little skull and its mandible *Australopithecus africanus:* the ape-man of southern Africa. He argued forcefully that Taung was human and had possessed human faculties, including language. Then Dart sat back and waited for the reviews and accolades to pour in.

In early February 1925, shortly after the publication of Dart's report, a subsequent issue of *Nature* featured commentary on the find by four British scholars. *Nature* is published in London; its review board consisted mainly of eminent British scholars. While all four commended Dart on his discovery of an extraordinary primate, they were skeptical of Dart's claims about the humanness of the fossil. Sir Arthur Keith of the British Museum pointed out that skulls of juvenile apes tend to resemble those of humans more than adult ape skulls do, and he cautioned the scientific community not to use the Taung child to revise our family tree (Keith later stated flatly that Dart's claim for Taung's humanity was preposterous). Grafton Elliot Smith, who had been Dart's teacher at University College, London, and had inspired him to pursue a career in anatomy, also cautioned against assuming that Taung was human. And Sir Arthur Smith Woodward, another scholar at the British Museum, said that the Taung fossil did not alter his view that humanity had its origins in Asia, not Africa.

Raymond Dart had not realized that the world would need much time to accept the true significance of the Taung child. No one but Dart had studied the fossil, and the scientific establishment perceived that this little-known anatomist laboring in the boondocks was rushing to judgment in submitting his results so hastily. Dart's account had appeared in a newspaper before it appeared in *Nature*, a move that Keith and company considered ethically suspect. Their rejection of Dart's conclusions also relied on more than a hint of Eurocentric racism. At the turn of the twentieth century, few early human fossils had been uncovered, and those that had been found came from either continental Europe or the Far East. Intense cultural and national pride hinged on which country's terrain would produce the earliest human. The British badly wanted that honor. For one thing, if early humans weren't British, why, they might be German or French — neither of which was an appealing thought to British science. And surely the origins of humanity could not be found in Africa, which Europeans regarded as populated by dark-skinned, technologically backward people.

Another reason for skepticism concerned the Taung specimen itself. Dart's paper clearly stated that the child was an *upright* walker with a modest ape-sized brain. This starkly contradicted the orthodoxy of the day. Thirty-four years earlier a young Dutch scholar and fossil hunter had found, against all odds, the first *Homo erectus* skull. Eugene Dubois had earned a medical degree but bore a lifelong obsession with fossils. He took up a career in anatomy but left his university post to sail to Java in the Dutch East Indies as a military medical officer so that he could follow his quest for human origins. Java should have been the last place on earth to look for, and the last place to find, a human fossil. Asia, as we know now, was not the cradle of humankind that many believed it to be. Besides, the tropical humidity of Indonesia makes fossil preservation unlikely. But Dubois had a fiercely independent nature that bordered on eccentricity, and this served him well.

Dubois arrived in Indonesia in 1888 and was soon spending all his spare time looking for fossils and all his spare money hiring others to do the searching for him. Eventually, he obtained a leave from his post and headed off to search for fossils with a team of workers. Starting on Sumatra but later moving his search to Java, Dubois toiled in futility for many months, until one day his team found a prize in a riverbank near the village of Trinil, on the Solo River. He and his workers excavated first a skull, then assorted other bones, from the muddy soil. All looked strikingly modern. The leg bones showed clearly that the fossil had stood fully upright, prompting Dubois to bestow the species name *Anthropithecus* (later changed to *Pithecanthropus*, then still later *Homo*) *erectus*.

In 1892, Dubois announced his find to the world. Java man, as the fossil came to be called, seemed to confirm the view that humanity's roots were in the East, the cradle of much of early civilization. The skull was not entirely like ours. It was flat on top and had a low sloping forehead and a strong, even brutish, appearance. But it was clearly more human than ape and was as good a candidate for the key to humanity's past as had ever been found.

Dubois encountered his own problems. Few scholars initially accepted the claim of humanity for the fossil. Dubois and his workers had been a bit sloppy in documenting the find, and critics seized on this to ridicule the idea that Java man represented a missing link. Critics called into question Dubois' training, or lack of it, in human fossil study. Nationalism also played a role, as French, British, and German scientists quickly rejected the discovery and the reputation of its discoverer. And, like Dart, Dubois had published his results quickly and perhaps without proper caution. Dubois became embittered when he found no eager scientific audience for his work and responded by withdrawing from what had become a debate. Another scholar, the German Gustav Schwalbe, obtained a cast of the skull and wrote his own lengthy research piece about it, agreeing with but overshadowing Dubois' controversial interpretations. The skull ended up as Dubois' personal Waterloo. His work received nothing but rejection and derision throughout the next decade, and the bones became dark secrets for much of the early twentieth century, locked away in Dubois' home, unavailable for study by other scholars. The man became a recluse as he grew older, refusing to see visitors who wanted to talk with him about the fossil.

The view that the first human was a big-brained quadruped — a brainy gorilla or chimp clambering about in the trees — had deep roots among European academics. The theory had received a big boost from the 1911 discovery of a fossil skull in an English gravel quarry called Piltdown. Picked up by Charles Dawson, a weekend fossil collector, Piltdown man not only embodied the big-brained ancestor that science considered likeliest at the time but had turned up on English soil.

Piltdown occupied pride of place on every scholar's genealogy of humans throughout the first half of the twentieth century. Experts had some initial consternation about the possibility that the pieces of the skull and jaw did not come from the same creature. But in 1917 more fragments were discovered, linking the mandible and skull perfectly. For twenty years every discovery of human origins, including that of the Taung child, was interpreted in the light of Piltdown man. Sir Arthur Keith was Piltdown's major booster among British scientists and promoted the idea that Java man and Piltdown formed an evolutionary line that linked apelike humans with later British forms.

Finally, as more and more fossil discoveries pointed to Africa as the cradle, and to a bipedal chimplike creature as the cradle's progeny, Piltdown man came to be seen as an anomaly. By midcentury it was clear that something was wrong with the place accorded Piltdown. Then chemical methods of dating fossils were devised, and Piltdown's true identity was revealed: The jaw was that of an orangutan, with points of attachment to the skull cleverly broken off, placed in the soil near a modern human skull. A clever prankster had planted the bone fragments, stained to appear ancient, among human bone fragments. For good measure, the fossilized bones of assorted other ancient animals were scattered with them. The perpetrator's identity has never been proved; for many years suspicion fell on Dawson, the discoverer, or Pierre Teilhard de Chardin, the theologian and philosopher who was also an amateur archaeologist. One theory even implicated Sir Arthur Conan Doyle, of Sherlock Holmes fame. Doyle fancied himself an intellectual and a scientist but had long felt mocked by the community of professional scientists, which regarded him as a mere novelist. Whoever the culprit was, acceptance of Piltdown had been quick and unanimous, because the preconception was so strong. Not only the public but Keith himself bought the false bill of sale.

Although the story of this infamous hoax has been told many times, it gives a good sense of why Raymond Dart's Taung child was not accepted for many years, although it was the key to understanding the real evidence about how we became human. Dart was finally vindicated when, in the 1930s, other South African australopithecines were unearthed, many by his friend and colleague Robert Broom, a Scottish physician and fossil hunter. Dart, who lived until 1988, never wavered in the face of hidebound thinking by politically powerful academic enemies. In the 1960s scientists finally recognized Africa as a center of ancient human activity rather than a backwater. Dart eventually wrote books about Taung and early humanity, and his ideas became the basis for a series of best-selling popular books by Robert Ardrey in the 1960s and 1970s.

In the mid-1980s, just as I was beginning graduate school, I attended the opening of the "Ancestors" exhibit at the American Museum of Natural History in New York. For the first time the most important human fossil remains — the crown jewels of the ancient world — were collected in one place and exhibited to the public. As an undergraduate, I had read about Dart's lifelong quest for acceptance and had seen photos of the Taung skull. In New York, in the vast museum auditorium, a small man with a good measure of antiquity himself rose slowly to his feet to accept thundering applause from the audience. Raymond Dart, then in his nineties, stood and accepted the praise, acknowledging a debt long overdue.

WALKING THROUGH HISTORY

When we want to find the roots of our thinking about evolution, we look to Charles Darwin. His were the first cogent writings about upright posture since Aristotle, who had referred to humans as featherless bipeds and to the act of walking upright as "that most basic of all habits," mainly to contrast it with loftier aspects of human uniqueness, such as morality and virtuousness.

Darwin not only founded the theory of evolution of life by natural selection, he also wrote extensively about human origins. In 1871 he wrote, "If it be an advantage to man to stand firmly on his feet and to have his hands and arms free . . . then I can see no reason why it should not have been advantageous to the progenitors of man to have become more erect or bipedal."

Darwin fixated on the human brain as the benchmark of humanity. He believed that the brain, bipedalism, and tool use were inextricably linked. Becoming upright, he decided, allowed our newly freed hands to make and use tools, which in turn placed an evolutionary premium on being a clever craftsman. The effects on humanity cascaded from that point. Standing upright would have allowed the invention not only of tools but of weapons. Using weapons would have allowed smaller teeth for chewing food to replace the impressive canines that early humans used to rip flesh in battle. This shift, Darwin argued, would have reduced the size of the jaw muscles and the jaws themselves, bringing the human skull into the form it possesses today. But Darwin could not know that millions of years of evolution separated bipedal posture, tool use, and expansion of the brain. The fossil record was nearly nonexistent when he wrote, and ascertaining the ages of fossils was impossible.

We must appreciate just how little Darwin knew about the evidence for human origins. Genes were unknown. But despite the paltry groundwork that had been laid for him, with cautious reasoning and brilliant insights, Darwin put together the puzzle of how humans evolved. While many of his contemporaries believed that brain size corresponded to intellect — the old racist pseudoscience of craniometry — Darwin categorically rejected the notion that brain size predicted intelligence in modern people. He pointed out that Neandertals — the only early human known at that time — had larger cranial vaults, therefore larger brains, than fully modern humans. Therefore, he reasoned, modest upgrades in brain size and organization don't necessarily make a fully modern person.

The famed German evolutionist Ernst Haeckel agreed with Darwin that walking upright was a more basic adaptation than brain size and staunchly advocated the idea. Haeckel created the name *Pithecanthropus alalus* (the speechless ape-man) to label a yet-unknown but theoretical human ancestor. Dubois initially adopted the first part of that name years later for Java man, although its name was subsequently revised to *Homo erectus*.

Five years after Darwin's pronouncements on bipedalism, Friedrich Engels, the social theorist who was Karl Marx's colleague, wrote extensively about the essential connection between standing upright and becoming human. Whereas Marx had stressed the continuum between humans and other animals (Marx had fleetingly considered dedicating *Das Kapital* to Darwin, who wasn't flattered), Engels sought the bold divide and found it in the concept of labor. In *The Part Played by Labour in the Transition from Ape to Man* (1896), he wrote that early humans had extended the process that other animals had begun. Instead of using their body parts as weapons or tools, our ancestors created tools as extensions of themselves. Uprightness allowed for these innovations, Engels wrote.

Engels was actually closer to the mark than Darwin on the timing of this change. Using modern radiometric dating, we know that the rapid expansion of the human brain didn't begin until millions of years after bipedalism arose and long after the first evidence of stone tool use. But most evolutionary thinkers focused only on the importance of the brain's role in shaping humankind. From the famed embryologist Karl von Bauer in the early nineteenth century through the great British anatomist Grafton Elliot Smith in the early twentieth, the brain seemed to be the key, evolutionarily. That most scholars who were building theories about human origins would turn to the brain seems obvious. It is, after all, the seat of cognition, language, and so much else associated with being human.

That perspective may continue to receive more attention than other theories, simply because it appeals to our addiction to drama and narrative. Misia Landau of Harvard Medical School has suggested that the story of a big-brained biped, armed only with his wits, is irresistible. It portrays our ancestor as a heroic figure who conquered great forces of darkness — savage predators, dangerous prey animals, and a harsh climate — to evolve into us. Against all odds, a maladapted David survived by defeating a host of Goliaths.

But the underlying truth is that early hominids would not have survived to extend their genetic lineage for millions of years unless they were extremely good at what they did, in terms of both their physical and mental prowess. Hardly the shuffling, defenseless creatures of our evolution myths, the australopithecines must have been much like the chimpanzees that I study today: tough, resilient animals able to adapt to a wide variety of habitats, climates, and food supplies. Australopithecines may have been talented climbers as well as marathon walkers, and they probably had the benefit of group cooperation. Throw in a few other skills, like rock throwing or stick brandishing, and the critter that emerges can drive a lion off its kill as effectively as any toothy carnivore could hope to.

The understanding of what it means to be bipedal did not really advance until the turn of the twentieth century. Sir Arthur Keith, writing ten years before his plunge into the Piltdown debacle, published a tome in 1903 that laid out his version of the anatomical roots of humanity. Keith followed the great tradition of scholars who studied the relationship between structure and function in the body by dissecting wild primates. Abetted by his research, Keith presented a view of human origins that was strikingly rigorous and modern for the era. Like others who followed him, he found great import in the anatomy and behavior of gibbons in Southeast Asia. He based his theory of a several-stage evolutionary progression to upright posture on a humble four-legged ancient monkey that evolved into a treetop-dwelling arm swinger like a gibbon.

Although not great apes — minor anatomical differences have relegated them to the group known as lesser apes — gibbons are among our closest relatives. They have barrel-shaped chests, big brains, no tail, and a shoulder anatomy that allows their upper arms to rotate completely around in their joint with the shoulder. This last feature is a fundamental hallmark of being an ape. Its presence in gibbons, along with their more primitive anatomy in other respects, led Keith and many who followed him to consider these small graceful apes to be the best examples of how a quadruped can evolve into a biped.

My only opportunity to watch wild gibbons came during the 1980s when I was conducting research on other primates in the forests of Bangladesh. The dawn broke with a distant whooping — hoolock gibbon families singing back and forth to each other across miles of forest. As if staking territorial claims in such a resonant way wasn't enough, these apes careened through the trees in the best Tarzan style imaginable. Watching them cavort overhead was hard on the neck but easy on the eyes; they are marvels of what natural selection can create. But the arm swinging is really more adapted to hanging under tree limbs to feed than for rapid travel. With their graceful long arms gibbons can reach fruits growing at the slender tips of branches by suspending themselves beneath them. Many anatomists believe that the same shoulder anatomy was instrumental in promoting an arboreal way of life in our earliest pre-bipedal ancestor.

But if Keith was correct, evolution may then have taken an

interesting turn. According to Keith, arm hanging was the key trait shared by humans and apes, and the last prebipedal ape was an avowed arm swinger. The shift from an arm-hanging ape to a bipedal hominid necessitated changes only to the lower limbs. Life in the trees may have even facilitated our ape ancestors' eventual bipedal life on the ground. Keith fiercely advocated the gibbon model in the first three decades of the twentieth century. He and a few other experts saw anatomical adaptations in gibbons today that they believed were also present in the earliest hominid. That hominid would have been an arm-swinging creature rather than a large-bodied, ground-walking ape such as a chimpanzee.

An all-star team of other evolutionary thinkers opposed the gibbon school. The opposition stemmed mainly from the deeply held belief in the obvious: The chimpanzee and gorilla more closely resemble humans anatomically and behaviorally. These experts decided that the clues to bipedalism's beginnings were more likely to be found in the great apes. Scholars ranging from Marcel Boule, the great French anatomist and describer of Neandertals, to the American paleontologist Henry Fairfield Osborn supported the chimpanzee model of bipedal roots. The chimpanzee "knuckle-walks," placing its forward body weight on one set of knuckles curled under the hand and so moves in an odd, modified form of quadrupedalism. This places the chimp on the ground and elevates it slightly above conventional four-legged walking. Large body size, the accompanying weight, and knuckle-walking might have pushed apes out of the trees originally. These ideas, combined with our closer resemblance to chimpanzees, made a persuasive argument for evolutionary thinkers in the knuckle-walking school. This school of thought was often referred to as the "troglodytian" school, for *Pan troglodytes*, the scientific name of the chimpanzee. To be an avowed troglodyte might in other circumstances not seem desirable, but in this academic context it was one strongly held point of view. The knuckle-walking model held sway through the middle of the twentieth century. The British anatomist William Le Gros Clark adopted it and, by teaching and by example, persuaded others to follow. In time, Keith himself dropped his arm-hanging model in favor of the knuckle-walking model.

In the 1940s a fiery young anthropologist and anatomist, Sherwood Washburn, entered the arena of debate. Washburn acquired a strong bias in favor of the knuckle-walking model while a student of Earnest Hooten's at Harvard. Washburn had honed his views as a graduate student in the 1930s, working as a grunt research assistant on one of the last great expeditions in search of wild primates in Southeast Asia. His supervisors wanted to know more about the behavior of wild primates (Hooten wrote the first scientific description of the bonobo, or pygmy chimpanzee) and how the primate body functioned in the wild, because the anthropologists hoped to use that information to understand human-ape comparative anatomy.

Washburn's goal was to understand how a primate's body was adapted to life in the rain forest by dissecting it on the spot. When the Thai field assistants carried into camp each day a dozen carcasses of gibbons that they had just shot (this was acceptable conduct in those early days of research), Washburn cut the apes open and studied their shoulder joints, hands, and feet. This field dissection, albeit gruesome by today's standards, gave Washburn a better appreciation than any previous student of primate evolution for the functional aspects of the primate body. He saw, for example, how the gibbon's rotating shoulder allowed seamless swiveling as it swung through the forest canopy. Washburn's fieldwork, combined with a razor-sharp mind and knack for synthesizing new disciplines, led him to the top ranks of evolutionary anthropology. He became the most prominent proponent of the knuckle-walking theory of human origins. In the generalized ape anatomy, Washburn saw a creature that natural selection could easily have tweaked to shift to upright posture. For carrying tools or other objects, and for going from place to place, Washburn saw knuckle-walking as the way our immediate prehominid ancestors must have traveled. Washburn became not only the standard-bearer for the knuckle-walking theory but also the dean of modern evolutionary anthropology. By merging sciences such as functional anatomy, genetics, and ecology, Washburn essentially created a new field of science during the 1950s and 1960s: physical anthropology, which is today known as biological anthropology. He then trained dozens of students in his worldview and sent them off to be the first anthropologists to study monkeys and apes as a way to understand early human behavior.

Still, debates about our origins have continued. Although Washburn fervently hoped that evidence of a fossil knuckle-walker would be found, other prominent scientists saw little evidence for knuckle-walking in our ancestry. For example, Russell Tuttle, a former student of Washburn's and now a prominent University of Chicago anthropologist, did extensive work in the 1960s on the hands of various ape species and found little support for the knuckle-walking idea. He has argued forcefully for a recent history of climbing and arm suspension.

Whether a ground-living or tree-climbing ape is the mother of us all may seem awfully esoteric. A great deal is at stake, however. The way our ancestors lived informs us about who we are now and about our past. The debate about human origins and the emergence of bipedalism has hinged on the discovery and interpretation of a number of exciting but enigmatic fossils, but gaping holes have remained in our understanding of our bipedal origins. To begin to solve the puzzle, we must first understand the way that apes live and move.

2
KNUCKLING UNDER

THE DAY IS SCORCHING HOT. The golden grass has a burnt odor, the palm fronds overhead are hanging limply, and I'm caught in a traffic jam. I'm not on an L.A. freeway but a narrow dirt trail winding through grassy hills in Tanzania. The traffic is a foraging party of chimpanzees strung out in front of me, pushing their way to a stand of fruit trees growing high on the slopes above us. The hill is so steep that, climbing behind the last chimp in line, my face is level with his rump. By the time we reach the ridge, I am hyperventilating, praying that they stop to rest, knowing that I will lose them if they turn upward again or take a difficult route to the top. To my relief I arrive at the hilltop to find the chimps gorging on a burgeoning crop of *Uapaca* — a fruit that grows only on the high ridges, which means a long haul for both the apes and the researchers who follow them. This is an annual routine in August and September, at the end of a long dry season, when the *Uapaca* become ripe. We're sitting on Bald Soko, a grassy hilltop so named in the early days of Jane Goodall's work here in Gombe National Park because researchers thought it resembled a balding chimpanzee.

After an hour of feeding, by which time I've caught my breath and am enjoying the spectacular view of the turquoise blotch of Lake Tanganyika in the distance, the chimps are on the move again. They head south, ignoring the trails made by humans in favor of dense thickets through which they gracefully push their way by using their forebodies. I am left to slither my way through thorny patches that grab me while allowing the apes to pass through unscathed.

By midday we have traveled about two miles. This is looking like a big travel day for the chimps, which is bad news for the researchers trying to keep up with them. The chimps knuckle-walk at about the same pace that I walk behind them. The difference is that they don't change their pace on achingly steep hills, whereas I slow to a crawl. Another difference is that after a day or two of long travel distances like today's, the chimps typically have a day or two of very little travel; they dawdle in one fruit tree rather than spend energy trying to find the next one. I mentally chalk this up to the problem of walking on your knuckles for mile after mile. But my disadvantage as an upright walker is that I am several feet too tall to slip easily through the thorny thickets.

A female chimpanzee knuckle-walks while carrying her daughter.

By late afternoon the chimpanzees have reached the southern terminus of their territory. Beyond stretches the range of the southern chimpanzee community; we are in no-chimps'-land, a place both communities claim as their own. This means that an ambush could occur at any time. But the chimpanzees push on, with males in front walking single file. By my estimate we have come nearly five miles since daybreak. The chimps eventually reach the beach of Lake Tanganyika before turning around and heading back into their home territory, where they bed down for the night.

In many forests in Africa chimpanzees and gorillas travel miles every day, and they do so in similar ways. Extending one arm far in front of their bodies, they simultaneously pull and, with their short legs, push themselves along. Watching them do this gives a perspective of how great apes travel that is far different from our popular image of their swinging through trees. And watching them also provides a far better perspective of humankind's humble roots than we could gain by studying the fossil record alone.

A fundamental difference between the body plan of a monkey and that of an ape lies in the range of motion for which their bodies are designed. Despite our image of monkeys swinging through trees, they do no such thing; instead, they run on four legs along the tops of tree limbs, leaping between the branches of tree canopies. Monkeys that spend substantial time on the ground, such as baboons, use the same skills to walk or run for hours on flat ground. If you look at the skeleton of a baboon or vervet monkey, you'll see an obvious resemblance to a dog or cat in all areas behind the skull. The torso and rib cage are deep and narrow, and the shoulder blades cover upper arms like pieces of body armor. And, unlike the apes, monkeys walk flat on the palms of their hands and feet.

An ape's body plan is utterly different. His rib cage is barrel-shaped, his shoulder blades pushed behind the arms to lie flat on the back of his shoulders. This keeps them out of the way of the shoulder. Where the humerus — the long bone of the upper arm — meets the shoulder, a complex joint and sac of ligaments allow the arm a full range of swiveling motion. All modern apes possess a rotating shoulder that allows them to swing from one arm to the other from the branch of a tree. This, along with its relatively larger brain, hearty chest, and lack of a tail, is what distinguishes apes from monkeys. In many ways apes differ more from monkeys than humans do from apes. The next time you see a chimpanzee in a zoo or on television, think about how insulting it is to the ape to be casually called a monkey — far worse than calling your human friend an ape.

The rotating shoulder that evolved in our ape ancestors allows gymnasts to swing on the high bar and has enabled Roger Clemens to throw 90 mph fastballs. Contrary to every cartoon de-

Our own shoulders rotate because they allowed apes to do this: hang by an arm from one branch while feeding from another.

piction, the rotating shoulder is not designed mainly for travel through treetops. Instead it allows an ape to hang beneath a tree limb, a useful position because most fruits ripen first at the tips of the smaller branches in a tree's crown. If a hungry ape traverses the branch from below, the wood will bend but not break, allowing the animal to reach its bounty.

But most apes spend relatively little time traveling in the treetops. The three African apes — chimpanzee, bonobo, and gorilla — travel from one place to another by knuckle-walking on the forest floor. A gorilla stands with its forebody elevated well above its rump and hindquarters, its weight suspended over its massive shoulders and arms. The hands are on the ground, four of the fingers curled under the palm. The only part of the hand that contacts the ground is the second row of knuckles. To prevent the hand from collapsing or hyperextending during knuckle-walking, the end of the apes' radius bone, which terminates in the wrist, has a ridge that limits how far the wrist can flex, so that the hand is more stable during knuckle-walking. It creates, in effect, a locking mechanism for the ape's wrist.

APE ANCESTORS

Apes weren't always knuckle-walkers. The first apes appeared on earth roughly twenty million years ago, as the forests of Africa began to recede in the face of a worldwide drying and cooling trend. The great savannas of East Africa began to cut into the expanses of tropical African forest. The apes diversified rapidly into many ecological niches and quickly vanquished *their* competitors, the monkeys. If you took a stroll through an African forest eighteen million years ago, its primates would look dramatically different from those found in the same place today. It was the golden age of the apes. In their diversity and wide range of ancient habitats, they dramatically outnumbered monkeys.

You might not have recognized these primates as apes, however. Some would have been as small as a few pounds. The earliest apes did not knuckle-walk or hang by their arms; they walked flat-footed on the palms of their hands and feet as monkeys do. Almost twenty million years of natural selection have substantially rearranged their body plan, crucial joints, ligaments, and perhaps even their internal organs. Scientists use fossil ape teeth to identify these progenitors of modern chimpanzees and gorillas. All African and Asian monkeys share the same tooth pattern. Each molar is topped by a pair of parallel high crests, a feature found only in these simians. These may be an aid in biting through tough leafy plants. The top of an extinct ape's molar lacks these ridges; it looks much like that of a modern ape — or yours or mine. Five bumps, which your dentist calls cusps, are connected to each other by fissures formed in the shape of the letter *Y*. This dental arrangement is unfailingly characteristic of any ape, regardless of what its feet and hands looked like. The earliest so-called dental apes walked with their feet and hands flat on the ground, just like any monkey.

We think that knuckle-walking evolved only once and became the ancestral posture and style for a wide assortment of great apes. The once-only assumption, which has few detractors, is based on what is often called Occam's Law, or the Law of Parsimony. Because bipedalism is a rare suite of anatomical traits in the animal kingdom, that it would evolve twice independently in the same lineage is about as likely as lightning striking your house twice. When presented with an array of related fossils, all showing signs of upright posture, it makes good sense to argue that the similarity is the result of shared evolutionary history, not a reinvention of the same wheel. Of course, if you live in a house struck once by lightning this is not a reassuring chain of logic in a thunderstorm. Each thunderstorm and each evolutionary chapter in our history is unique, and only magical thinking can persuade us otherwise. One such second lightning strike may be *Oreopithecus*, whose case we will take up later.

The best known of the dental apes is a creature named *Proconsul*, which was discovered in Kenya three years after Raymond Dart's discovery of the Taung child. The fossil was named after Consul, which was a common name for performing chimpanzees of the day, just as Rover was used to name dogs. Actually a group of several similar species, *Proconsul* was a medium-sized ape that lived in the trees of African forests between twenty and seventeen million years ago. Like the other dental apes, this one did not look much like an ape. If you could watch a *Proconsul* running around a zoo enclosure, you would probably misidentify it as a monkey. *Proconsul* did not knuckle-walk, nor did it climb as expertly as a modern chimpanzee. Until recently, *Proconsul* held a heralded place in our own evolutionary history: It was considered the last common ancestor of apes and humans.

In 1932 a Yale University graduate student named George Edward Lewis was searching for human fossils far from the African cradle that Dart and others had begun to champion. Lewis was working in the hills of northern British India, in what is today Pakistan, where he found a broken bit of the upper jaw of a very primitive ape-human (some say Lewis actually purchased the jaw from a local fossil vendor). The jaw had some teeth remaining, all of them smallish, with thickly enameled molars, and was believed to date to about fifteen million years ago. Lewis bestowed the honorific name *Ramapithecus* on the specimen, naming the ape after the hero of the Hindu epic the Ramayana.

That fossil went through infamous interpretations. Lewis interpreted the bit of bone as hominid, much to the consternation of his colleagues. He deposited the fossil in the drawers of the Peabody Museum at Yale, where it collected dust for thirty years. In the early 1960s another Yale fossil expert, Elwyn Simons, took renewed interest in *Ramapithecus*. He and his graduate student David Pilbeam undertook a study of the *Ramapithecus* specimen and concluded that, despite the absence of anything more than a partial jaw fragment, this was the direct ancestor of modern humans.

Eighteen million years ago a forest in Kenya held a diverse community of apes. The apes at upper left and on the ground are two depictions of *Proconsul.*

For the next fifteen years, with little evidence and despite the heated disagreement of their colleagues, the pair claimed human status for the fossil. *Ramapithecus* was the Lucy (the Rosetta stone of all early human fossils) of its time, depicted in textbooks as walking upright, even carrying primitive tools.

Unfortunately for Simons and Pilbeam, at about the same time biochemists were developing tests that could gauge the degree of immunological distance between humans, chimpanzees, and other primate relatives. What's more, Vincent Sarich and Alan Wilson of the University of California, Berkeley, announced in 1967 that they had devised an immunological clock that established the date of divergence among modern apes and ourselves. That date was only five million years ago, astoundingly recent compared to the conventional wisdom at the time and far too recent for the *Ramapithecus* fossil to have been a direct human ancestor. In fact, Sarich at one point declared flatly that *Ramapithecus* could not possibly be human, because it could not have been an upright walker.

Although Sarich and Wilson initially infuriated the paleontological establishment, it came around to Sarich and Wilson's point of view later, especially when Pilbeam himself excavated more *Ramapithecus* specimens in Pakistan that were clearly more ape than human. Today we consider *Ramapithecus* and its close relative *Sivapithecus* to be likely ancestors of later Asian great apes, perhaps including the modern orangutan.

The paleoanthropologists Daniel Gebo of the University of Illinois and Laura MacLatchy of Boston University identified another fossil ape in 1997, after they re-examined fossil fragments from Uganda that had lain dormant in a museum drawer for years. The researchers, who also examined fossils excavated at the same site in 1994, recognized the importance of the Uganda fragments. They named the specimen *Morotopithecus bishopi;* its fragments point to a chimpanzee-like ape living twenty million years ago that was endowed with the arm-hanging shoulder apparatus that the dental apes lacked. *Morotopithecus* is, at least for now, the wearer of the mantle of the last common ancestor of both apes and humans.

Later fossil apes began to make the transition from a climbing quadruped to an arm-hanging one. By ten million years ago we find creatures such as *Oreopithecus*, affectionately nicknamed "Cookie Monster" by generations of anthropology graduate students. *Oreopithecus* was an ape of the Mediterranean region that possessed a chimpanzee-like arm-hanging shoulder anatomy. A related fossil ape, *Dryopithecus*, was recently found near Barcelona and appears to have had the same features. We have found few good fossil ape or protohuman specimens for the period between twelve and five million years ago; the remains are mainly teeth and jaws, offering little evidence of how the creatures moved. Many researchers have therefore turned to models based on what modern apes do and don't do with their hands and feet. But this has not fully resolved the issue.

A fossil ape as its skeleton would have moved in life.

Of the myriad fossil apes that have been discovered, knuckle-walkers and otherwise, one more merits special attention. During the 1930s the fossil hunter Ralph von Koenigswald searched eastern Asia from China to Indonesia for evidence of fossil humans. In addition to field excavation, von Koenigswald searched drugstores, because Asian people frequently attach strong traditional beliefs to the powers of bones of various animals, which they grind and eat as medicine or aphrodisiacs. Von Koenigwald's hunch paid off in 1935 in an apothecary shop in the Philippines. He purchased several humanlike teeth and then in Hong Kong found several hundred more. In this collection he found one immense molar that was from an ape or human species unknown to science. He named the creature *Gigantopithecus.* Based on the molar size, the creature would have been about twice the size of a silverback mountain gorilla and weighed seven hundred pounds or more. In the wake of the widespread derision of Dart's Taung child, *Gigantopithecus* was for a time accepted as a possible immediate human ancestor, based only on a tooth. Further remains of the giant primate were found on several expeditions all across Asia in the ensuing decades, and the identification of *Gigantopithecus* became clear. It was not one but several species of enormous ape and probably represented a side shoot of the ancestral branch that went its own way and ultimately faded into extinction. But some believe that the legends from Asia of a yeti, or abominable snowman, and those from the Pacific Northwest of Bigfoot, arose eons ago from encounters between earlier generations of modern people and *Gigantopithecus.*

WRIST ACTION

You might think that walking on four legs would be a more logical precursor to walking upright than climbing vertically, but most researchers feel the opposite is the case. Being an adept climber preadapts the body for a later transition to bipedal walking.

Climbing requires longer arms than legs, a torso that can assume a vertical posture, and grasping hands and feet with long curved fingers and toes. Most of all, habitual climbers must be able to move in multiple dimensions while ascending trees. A climber must have powerful but highly mobile joints and a shoulder that allows a full range of arm motion. An animal that makes its living by running or walking on flat ground needs only a shoulder joint that holds the arm in place while it swings rapidly in a semiarc.

In papers published in the 1970s the anatomist Jack Stern and his colleagues of the State University of New York at Stony Brook argued persuasively that the similarities of the ape and human thorax — the body's midsection — and upper limbs come from ancestors who assumed an upright climbing posture when in the trees. This became a modern revision of Keith's arm-hanging theory. Stern wrote that an ape that climbed upright was predisposed to walking upright on the ground too. Its long limbs would have been useless for propelling it on the ground, so the logical course for natural selection to take would have been to free the arms from all responsibility. Later, Stern and his associate John Fleagle claimed that the forelimb features of apes and humans that Keith had ascribed to arm hanging were better attributed to adaptations to climbing.

The debate has led some researchers to the hand rather than the feet for clues about how ancient apes walked. The unique ape wrist is at the center of the controversy. The bony features of the ape and human wrist have long been interpreted as evidence of an arm-hanging ancestry, which would support Keith's old theory. But Fleagle and his colleagues disagree, because chimpanzees are far more accomplished ground walkers than gibbons but possess wrists better adapted for arm swinging. Fleagle and Glenn Conroy of Washington University think that this is the smoking gun that points to a knuckle-walking ancestor as the first biped.

But Daniel Gebo of Northern Illinois University offered another view, one that backed Keith's idea that arm hanging led to bipedal posture. Gebo pointed out that evidence of vertical climbing appeared in the fossil record of Old World monkeys twenty-five million years before the rise of bipeds. Based on his studies of how modern primates move about in trees, he concluded that the earliest apes were probably four-legged climbers that evolved into arm hangers. These swingers in turn evolved into knuckle-walkers, agile both in trees and on the ground, and later became upright walkers. In the shared traits of the chimpanzee and human wrists, Gebo saw evidence of weight-bearing adaptations that suggest a ground-walking phase immediately preceding upright walking.

In short, the ancestral ape that gave rise to us all either lived in the trees and adopted knuckle-walking after the split with the human lineage, or came down from the trees and knuckle-walked before the split. The notion of a knuckle-walking phase in our direct ancestry got a big boost in 2000 from two George Washington University anthropologists, Brian Richmond and David Strait. They looked at an early hominid wrist and found, at the terminal end of the forearm's radius bone, remnants of the wrist-locking mechanism that exists in the wrist bones of a chimpanzee or gorilla. As the ape's radius makes its way to the wrist, a bony ridge flares up from its surface. This ridge, combined with an angle formed by a notch at the end of the radius, locks in place and limits the ape's wrist extension. Richmond and Strait felt that this was strong evidence that the last ancestor of the earliest humans was a knuckle-walker.

If verified, the bold claim of Richmond and Strait echoes the words of Sherwood Washburn in the 1940s and would overturn a half-century of orthodoxy about how we became human. But the study is not without some serious detractors, who have criticized the new study's authors for their methods. The five skeletal features in the wrist that the researchers used were recorded with video equipment, and some of the features were oriented for analysis by sight rather than with precision instruments. What's more, the conclusions were based on only two specimens of the early hominid wrist, one of which was that of the famed fossil Lucy. Lucy's wrist measurements actually didn't support the argument of Richmond and Strait nearly as well as another, more obscure specimen of the same species that was available only as a replica. Richmond and Strait's specimen may also have been missing some pieces that would have changed the analysis entirely. Although initially the study seemed to offer the answer to an old debate, it now appears that the question of whether hominids evolved from a knuckle-walker or a climber remains wide open. This sort of contention over bits of bone is common and essential in paleoanthropology.

WHAT APES DO, AND WHY

Unlike us, apes walk around all day on all fours, and their ancestors walked this way too. That walking with all four legs on the ground was an important part of the lives of ancestral apes makes much sense to anyone who has spent time following modern great apes around African forests. The African apes travel mainly on the ground, ascending into trees to search for fruit or to spend the night. Despite their long arms and curved fingers, chimpanzees and bonobos are well adapted to ground travel. Chimpanzees can travel several miles a day over rugged, thickly wooded terrain — not bad for creatures walking on their knuckles. They walk great distances to find ripe fruit, which is about 70 percent of the diet of both chimpanzees and the closely related bonobo. Understanding how this affects their behavior goes a long way toward explaining the benefits of bipedal walking. If apes were shown, for instance, to be inefficient when knuckle-walking, the advantages of bipedalism would be immediately apparent.

Unlike leafy foliage, the basis of the diet of many other primates, fruit is fleeting in availability and patchy in distribution. In tropical forests individual trees of one species, such as figs, may be widely scattered. To find figs chimpanzees must undertake a high-energy lifestyle: The calories needed to walk long distances to find fruit are paid back only by the carbohydrate content of the fruit itself. The energy comes with a high cost, that of trekking long distances. I have followed chimpanzees for hours during which they rarely paused until they reached a certain towering stand of fruit in a remote ravine. I arrived panting and aching, and the chimpanzees were already feasting on the rewards of their labor.

This focus on fruit exerts a profound influence on chimpanzee social behavior. A female chimpanzee's career revolves around finding the right quality and quantity of food to nourish herself and her offspring. When female chimpanzees migrate at puberty to other, neighboring community territories, they settle into a breeding area that must nourish them and their offspring for decades. Meanwhile, males travel farther and wider, walking faster than females do. Females do not travel in a large close-knit group with males and other females in the way that female gorillas and many other primates do. Instead, female chimpanzees travel alone or with just a few companions. When they find a tree laden with ripe fruit, they eat as much as they can with as little competition from others as possible.

Females become highly sociable at one time, however — when they are in estrus. The enormous fluid-filled pink swelling that balloons on a female's rump for a portion of every month is her billboard of availability to males, and when she is swollen she is more drawn to males as well. Most of the time, however, a female forages on her own, often with an infant clinging to her belly or riding on her back. This energy strain is no doubt why she is more sedentary and solitary than a male; females without infants travel just as far and fast as males. The higher cost of being female may account for chimpanzee society's consisting of sociable, wandering males and less sociable females.

Males are also fiercely territorial, regularly patrolling the periphery of their territory and monitoring intrusions by chimpanzees from neighboring communities. If they encounter lone strangers that have accidentally wandered into their territory, they attack and sometimes severely injure or kill them. Chimpanzees are the only primates other than humans to engage in this sort of lethal defense of territorial borders, another feature of their society that points to primitive aspects of our own nature.

Martin Muller, a graduate student of mine who has studied chimpanzees in Kibale National Park in Uganda, witnessed the immediate aftermath of one such intercommunity encounter. The male chimpanzees that he was following met a strange male, apparently out alone, as dusk was falling. Martin listened to the screams but had to return to camp for the night. The next morning he returned to the scene and found the strange male lying face up, dead, in a clearing of flattened vegetation that extended for yards around the body. The victim was brutally wounded, his body punctured numerous times, his windpipe torn out, his scrotum torn off. Rolling the victim over, Martin found that the back of the body was unscathed. The males of the home community had apparently pinned the intruder down while their compatriots savaged him.

Male chimpanzees also hunt other mammals. Unlike their cousins the bonobos, chimpanzees are avid predators. A community with ten males may kill hundreds of monkeys a year, plus occasional antelope fawns, wild pigs, and other, smaller prey. As chimpanzees knuckle-walk along the forest floor, they spy monkey groups clambering about in the treetops. Stopping to watch, the members of the foraging party decide whether to climb into the canopy to pursue the prey. The usual victim is small, just a few pounds or less, but frequent hunting may yield more than a thousand pounds of meat annually.

Females relish meat too, but they tend not to cooperate with one another to hunt, and males always monopolize the kill. Males use meat as a social currency of sorts, offering it to females and other males as a way of enhancing the status of the hunters. My research has shown that cooperation is more important in hunting success than having huge canine teeth or long climbing arms. One of the most successful hunters among the Gombe chimpanzees I studied was Evered, an elderly male who lacked nearly all his teeth. Evered was as skilled a hunter as any of his much younger, more muscled comrades, probably because his many years of hunting monkeys had endowed him with the experience that was a greater key to success than weaponry. My point here is simply that agile, tree-climbing apes do not need to walk upright at all to obtain meat. As we shall see later, however, once one has adapted to living on the ground, a whole new supply of meat becomes available.

Some scientists have put forward a close relative of the chimpanzee as the best model for bipedalism's origins and of what the earliest humans may have been like. The bonobo is similar enough to a chimpanzee that if we found the two as fossils in the same archaeological dig, we would no doubt assign them to the same species. But geography and behavior separate them. Bonobos probably evolved from a population of ancient chimpanzees cut off from the mainstream of chimpdom when the great Congo River, which cuts a broad parabola across Africa's midriff, changed course eons ago. Isolated genetically from other chimpanzees, protobonobos went their own way. Like chimpanzees, bonobos live in complex communal societies in which fruit is the staple. Females migrate between groups, and males defend community territories. But while even the lowliest adult male in a chimpanzee community can dominate all the females, female bonobos form coalitions that can dominate males or at least prevent males from brutalizing them. This power-sharing arrangement has served female bonobos well; they typically receive priority of access to food, including meat, when prey is caught.

The bonobo, a close relative of the chimpanzee, is often held up — incorrectly — as a better model for the origins of upright walking.

Female empowerment among bonobos has led researchers to hold them up as an alternative model for the origins of human behavior and gender values, in stark contrast to the brand of male domination that we see among chimpanzees. And, because of the way that bonobos walk, some researchers have argued that bonobos are the more appropriate model of what early humans may have been like. In the 1970s the primatologist Adrienne Zihlmann of the University of California at Santa Cruz and her colleagues proclaimed the bonobo more similar to humans in posture and gait than are chimpanzees. Zihlmann and colleagues based this claim on minor anatomical differences: Bonobos have arms and legs that are, relatively, the same length as a chimp's and a torso, upper body, and hips that are narrower. Numerous other researchers immediately challenged Zihlmann's assertion, pointing out that all the supposedly humanlike traits could be easily explained as adaptations to vertical climbing in trees.

The depiction of bonobos as more bipedal than chimpanzees, and therefore more appropriate as models for human origins, grew during the 1980s and 1990s. A television documentary featured many shots of female bonobos strolling around central African forests like hunter-gatherers. These shots had been artfully edited from the assortment of mostly quadrupedal footage that had been filmed. Wild bonobos walk upright occasionally, as do wild chimpanzees. But the idea that bonobos walk upright more often, or are somehow more bipedal than chimpanzees, is wrong.

Recent research by Elaine Videan and William McGrew of Miami University has shown that in zoos, bonobos are no more bipedal than chimpanzees. Videan and McGrew compared chimpanzees and bonobos of similar backgrounds — circus, zoo, the wild — and found that the tendency to stand upright was essentially the same for both species. In the wild little evidence exists for bonobo bipedalism, except in odd circumstances where the animals are fed by people and therefore walk upright while carrying handfuls of bananas or sugarcane. Diane Doran of the State University of New York at Stony Brook and Kevin Hunt of Indiana University compared their bonobo and chimpanzee data. They confirmed earlier findings that bonobos spend more time in trees than chimpanzees do and that male bonobos make much more use of their rotating shoulder than male chimpanzees do. Bonobos appear to be better adapted to life in the trees and chimpanzees better adapted to traveling on the ground. This supports the chimpanzee as the better model of the way early humans traveled in the days before bipedalism, but more bonobo populations must be examined before we reach a conclusion.

This is another in a long list of claims about whether bonobos or chimpanzees are the better model of our last common ancestor. The bonobo reached scientific daylight only in the 1980s, centuries after the chimpanzee was revealed to Western science and entered the public imagination. The first reports were about captive bonobos and described them as peaceful and hypersexual. One primatologist said that bonobos settle their disagreements by making love rather than war. He reported that the intensity, frequency, and variety of sexual behavior in bonobos were unrivaled by those of any mammal other than humans'. Among all primates, only human and bonobo females seem freed from the sexual constraints of estrus to have sex with males — and sometimes with other females — whenever and however they choose. The bonobo-chimpanzee dichotomy is partially true. Female bonobos truly display more power and alliance formation than female chimpanzees. Bonobos in zoos engage in a rich array of sexual behavior: male with males, females with females, and males with females. As one bonobo researcher told me, when they aren't mating, they're usually masturbating. But when field scientists compare the behavior of wild bonobos and wild chimpanzees, many of the contrasts disappear or are greatly diminished. The early report that female bonobos mate year-round apparently is true only in captivity. In the wild both chimpanzee and bonobo females have the vast majority of their sexual encounters when their sexual swellings are maximally engorged.

The last of the African apes is by far the most restricted to life on the ground, because of its anatomy and gargantuan size. Gorillas spend most of their time walking on the ground on all fours, which is not surprising given that some males can top 450 pounds. Their anatomy reflects this: The gorilla shoulder joint is less adapted to climbing than that of chimpanzees and bonobos. Most of us think of gorillas as lumbering giants browsing in high mountain meadows like cows, courtesy of Dian Fossey's work, which was memorialized in the book and film *Gorillas in the Mist.*

This is, however, something of a misrepresentation. Although mountain gorillas are in fact quite sedentary, bulldozing their way through ferny thickets and bamboo stands at a slow pace, other gorillas are not. Lowland gorillas, which compose most of gorilladom, conform more to a chimpanzee pattern of long-distance travel and ripe fruit eating.

In the Impenetrable Forest of Uganda, where I have been studying the ecology of both chimpanzees and gorillas since 1996, we watch gorillas climb into enormous trees to feed on fruit, fungi, and epiphytic plants that grow 150 feet overhead. I know of at least one gorilla, elsewhere in Africa, that fell to its death while foraging high in such a tree, but tree climbing nonetheless is a regular part of gorilla ecology. In forests where fruit is plentiful, gorillas follow the chimpanzee pattern, foraging far and wide to find it. Only in marginal habitats like the Virunga volcanoes, where elevations of 10,000 feet and cold misty conditions make fruit trees scarce, do gorillas subsist instead on fibrous ground-living plants. While in the trees gorillas climb in an upright posture, although they do stand or walk bipedally.

The relationship between gorillas and their locomotion in trees might seem simple: the bigger the body, the bigger the tree limb needed for support. But the relationship is not that straightforward. The primatologist Melissa Remis of Purdue University showed that female and immature lowland gorillas climb more than males. Male climbing was also influenced by their social group: When a male was left on the ground while his group foraged in trees, he climbed up too. Silverbacks rarely traveled from tree to tree directly; their bulk required them to descend carefully to the ground and amble across the forest floor to the next tree and climb it. Remis observed as much arm hanging by female gorillas as a chimpanzee watcher would see, although the gorillas tended to use their legs as well for an extra boost between limbs.

Gorillas are as generalist in their approach to walking as

chimpanzees are. They adapt to the local structure of the forest. Ferny ground thickets lead mountain gorillas to plod along on the ground, while a richer patchy food base leads lowlanders to do marathon walking each day. Although my team in Bwindi Impenetrable National Park in Uganda studies mountain gorillas that travel only about eight hundred yards a day on average, lowlanders in other forests may cover as much as four times that distance — and need extra time and calories to do it.

The only Asian great ape, the orangutan, is a habitual climber, rarely traveling on the ground and virtually never bipedal. They are "quadrumanous," adept at using their feet as third and fourth hands as they scramble about Indonesian rain forests. Orangutans come to the ground to cross forest gaps, their massive red-haired bulk supported by a strange variant of knuckle-walking

A great ape walks on its next-to-last set of knuckles. Note this chimpanzee's opposable big toe.

in which the weight of the forebody is borne on the outer edges of the fists. They look like they are about to shoot marbles as they amble about on the ground.

Presuming optimality would mislead us entirely about the orangutan. The huge size difference between males and females, and the orangutan's close links to the highly social gorilla or chimpanzee, would lead someone reconstructing orangutan behavior from its skeleton alone to argue for a group-living ape. Instead, the orangutan is largely solitary. Its recent ancestors lived not only in tropical rain forests but on the hills and plains of Asia from Pakistan to southern China; perhaps they led a life very different from that of today's orangutan and their anatomy has not had time to catch up with the change in society. In any case, the orangutan reminds us of why we must be careful when theorizing about the connection between ecology and behavior.

The environment exerts a tremendous influence on the evolution of great ape societies, providing the filter through which natural selection acts. We can study the fossilized bodies of ancient apes and early humans to get an appreciation of the forces that molded them and the way that they must have led their lives in response to those forces. But in reconstructing the behavior of animals from their fossils, we face a problem that has no solution. We don't know how tightly evolution has molded a species to its habitat. The very small number of living species of great apes, a tiny fraction of the total species diversity in the ancient past, further limits our ability to extrapolate. We should never assume that we will find our ancestor among the living species of apes, but they can point us in the right direction for the search.

3
HEAVEN'S GAIT?

THE UNDERGRADUATE class trip to the San Francisco Zoo was an annual ritual in the late 1980s when I was a graduate student instructor of human origins at the University of California, Berkeley. We led wide-eyed students past the gibbons, the gorillas, and the orangutans, always making a planned stop at a certain spot by the chimpanzee exhibit. I would commence a short lecture about chimpanzee behavior as the class alternately looked at the apes and me. The male chimpanzee sat like a Buddha on his rocky grotto, forty feet across a cement trough from us.

As I spoke, the Buddha would begin to stir. He would almost imperceptibly begin to rock his body to and fro, and a low-pitched moan would emanate from his throat. The students rarely noticed this movement until it was too late. The Buddha would suddenly raise himself upright and begin to stamp his feet madly, then rock from one foot to another in a two-legged standing position. Every hair bristled from the muscles of his aging body; it was quite a show, and the students were riveted. He would then surge about his rocky outcrop, picking up piles of his feces and flinging them with impressive accuracy at the students (I was always careful to take a step back as this unfolded) from his bipedal standing posture. If they remembered nothing else from that semester, my students always knew years later who had been "slimed" by that lonely, bored chimpanzee.

. . .

That chimpanzee had figured out a way to use upright posture, at least momentarily. We think of two-legged walking as an occasional, transitory behavior for chimpanzees, merely a way of getting from place to place. Although this is partly true, standing upright involves far more than that. It is first and foremost a dramatic change in posture involving a domino effect of far-reaching permutations in other parts of the body, and only secondarily is it a change in gait. Understanding the emergence of bipedalism is a mere guessing game unless we can evaluate its costs and benefits. Of posture and gait, only gait involves easily measurable costs and benefits. In understanding how, why, and when the shift from a quadruped to a biped occurred, gait represents a common denominator.

That common denominator is the principle of natural selection. The guiding force in evolutionary theory predicts that in each generation, ever-so-slight variants of each feature of an animal will be selected, or not, to be represented in the next generation. In this way tiny increments of change creep into a species, changing it from one thing to another or splitting it into several new species over many generations. Darwin, who offered the first explanation of how the process works, did not know about genes or mutations, the key ingredients for evolution by natural selection. Genes are the currency by which the environment sifts out physical traits. They are long, redundant segments of DNA, reshuffled each generation during the act of reproduction when the genetic template of each parent is cut into the deck of evolutionary cards. The genetic code of the offspring begins with the combined complement of each parent and can be altered by mutations that replace the biochemical bases of the DNA strand. Such mutations are, far from our science fiction-bred imaginations, commonplace and usually have no effect whatsoever. The vast majority of mutations are favorable or unfavorable depending only on the environmental context.

Although we have the notion that natural selection acts only over great spans of time, this is not necessarily the case. Peter and Rosemary Grant are biologists who have spent decades studying the same Galápagos finches that Darwin collected while serving as the naturalist on his famous sea voyage aboard the HMS *Beagle.* Their study site is a desolate rock, hardly big enough to be called an island, in the archipelago better known for its giant tortoises and seafaring iguanas. The Grants have been monitoring a population of finches that has endured intermittent drought and famine, part of the natural boom-and-bust cycle of life on islands utterly dependent on rainfall for their plant life — and the birds depend on the plant life. A bird that hatches with a slightly odd bill strong enough to crack open seeds twice as hard as any it is likely to encounter has gained no advantage in its quest to survive and reproduce. But the same overly thick bill, produced by a random mutation at a time and place where food is scarce, may allow the bird to crack open the only food remaining in times of famine. In those lean times our mutant finch capitalizes on its neighbors' ordinariness by eating better and therefore hatching more eggs and leaving more progeny in the next generation. A bird generation span is short, and just a few decades of hardship may breed an entirely new bill shape in the population; thick-billed birds, all descended from our mutant, come to predominate. The Grants have found that the shift of bill size and shape can occur over just a few finch generations of lean times. In easier, rainier decades the need for a big bill decreases, and the easing of natural selection pressure results in finches with smaller bills once again sweeping the islet.

Switch now from the bills of little birds to the body plan of an ancient ape. If an ape can prosper more than other apes by moving from one place to another on two legs instead of all four, natural selection provides the game plan for the evolution of a biped. But walking on two legs couldn't have happened in such simple fashion, just as a winged, feathered bird didn't hatch from an egg laid by a scaly reptile. In each generation natural selection had to favor the form that our protobiped took, without any plan for building an even better model. This idea is the hardest aspect of evolution for most of us to wrap our minds around; it seems so obvious that because apes knuckle-walk and we walk perfectly upright, everything in between was designed to improve from the former to the latter.

We like to think of ourselves as the apex of evolution. But unfortunately for our collective egos, nothing was inevitable about the direction that hominid evolution took nor was there a blueprint for how to get there. Had a protohominid not made good use of its quasi-bipedalism, natural selection would not have pushed its genes differentially through to the next generation. Nothing about natural selection is forward-looking, no matter how hard this idea is to accept.

THE HIGH COST OF TRAVEL

To understand why radical changes in posture and gait evolved, we must use some simple caloric math to estimate the energy required to carry a body from point A to point B. We can study the energy cost — calories expended — and benefits — the distance traveled — by having animals walk or run on a treadmill in a windowed chamber equipped to measure exhaled air. The researcher measures the airflow into and out of the box and records the change in oxygen concentration. The doyen of animal oxygen-consumption studies was the late Harvard University zoologist C. Richard Taylor. With numerous colleagues he collected information for more than two decades on a wide range of wild and domestic animals and provided grist for arguments about the merits of quadrupedal versus bipedal walking and running.

Remember two key factors when thinking about animal locomotion. One is locomotor efficiency, the ratio of calories expended to distance traveled. The second is locomotor economy, the thriftiness of energy output for a given physical task. A big animal can travel more efficiently than a small animal, because the former expends less energy to move each ounce of body weight a given distance. An elephant can walk the length of a football field more efficiently than a mouse. The elephant, however, burns far more energy making the trip than the mouse does because of the pachyderm's vastly greater bulk. So the elephant wins the race in energy efficiency, while the mouse wins in locomotor economy.

Taylor and his team found that, as you might expect, calories burned for every unit of distance rise in direct relation to speed of travel. Ask our elephant to gallop from goal line to goal line, and its energy output rises dramatically relative to strolling the same one hundred yards. This relationship was amazingly constant for most mammals and birds, from hedgehogs and emus to elands and partridges. But Taylor and others found a few notable exceptions. Kangaroos travel more efficiently while hopping at high speeds

Economy = Total cost for a given task

The difference between locomotor efficiency and locomotor economy.

than at low speeds. Recent research by Timothy Griffin of the University of California, Berkeley, and Rodger Kram of the University of Colorado (and a former protégé of Taylor's) showed that penguins use less energy per step than people do. Penguins have long been notable for wasting more energy while they waddle than does any other biped. Some penguin species were thought to violate the fundamental laws of locomotor efficiency by waddling a hundred miles or more across icy terrain to reach their rookeries. But executing their clumsy-looking side-to-side rocking waddle actually helps increase their locomotor efficiency by raising their center of gravity slightly, so their little leg muscles don't need to push quite so hard. What seemed to be a paradox is actually just another piece of evidence that bipeds are efficient walkers.

Walking efficiency generally comes down to simple equations. Whenever an animal tries to defy gravity by climbing, it pays for that effort. Arboreal animals have greater locomotor costs than those that travel on the ground. The trade-off may be that once the vertical travel is done and an animal is in the forest canopy, it can travel horizontally — by running or arm swinging — relatively efficiently. But a ground-living animal pays a steep price for its velocity. To move faster such an animal must either lengthen its stride or its rate of striding or both.

This brings us to two-legged walking. Taylor and V. J. Rowntree compared chimpanzees and humans for walking efficiency. They noted the long-standing debate about the relative merits of walking on four versus two legs and that other bipeds — notably the ostrichlike rhea — expend twice as much energy walking on their drumsticks than four-legged animals do. Taylor and Rowntree found, much to their surprise, that a chimpanzee walks upright just as efficiently as it walks on all fours. They found the same result for a capuchin monkey made to walk both ways as well. The two researchers concluded, in an article in the prestigious journal *Science* in 1973, that the argument of locomotor efficiency as a rationale for the evolution of upright walking should be thrown out.

All scientists share a deep skepticism of one another's results. Researchers don't believe counterintuitive findings unless they can scrutinize the results themselves. In 1984, the biological anthropologists Peter Rodman and Henry McHenry of the University of California at Davis re-examined Taylor and Rowntree's study and came to a different conclusion. Whereas Taylor and Rowntree had compared human and chimpanzee locomotor *efficiency*, Rodman and McHenry compared the *cost*. They found that humans are not necessarily more efficient walkers than all quadrupeds, but we certainly walk upright more efficiently than chimpanzees do. In other words, it would make no evolutionary sense for most four-legged mammals to evolve upright posture, but if an animal has already evolved into a knuckle-walker, as apes had, moving to a fully upright posture was the energetically sensible thing to do.

In the 1990s the zoologist Karen Steudel of the University of Wisconsin became a protagonist in the debate. The chimpanzees used in the original treadmill studies, Steudel pointed out, were only juveniles and therefore may have had a very different energy in–energy out balance than that of adults. Taylor had used young chimpanzees because persuading a little one to walk a treadmill is easier than convincing an irascible adult. In humans the locomotor efficiency of children is much lower than that of adults, so Steudel argued that Taylor and Rowntree's results might have been invalid in the first place. She reasoned that even though modern people walk more efficiently than any quadruped of the same size and weight, this wouldn't necessarily have been true of the earliest hominids that were not brand-new bipeds. Therefore, the initial motivation for moving into an upright posture and behavior could not have been energy efficiency.

Steudel also tried to turn Taylor's data against Rodman and McHenry: Taylor had used data by a separate research team that seemed to show that bipedalism is more expensive than quadrupedalism. Steudel cited this finding. In rebutting her, William Leonard and Marcia Robertson of the University of Florida pointed out that the study cited by Taylor and Rowntree examined *running* costs, not walking costs. At walking speeds, Leonard and Robertson found, as Rodman and McHenry had before them, that bipeds are simply more efficient movers of their own body weight than quadrupeds.

These studies did not gain widespread acceptance because of their tiny number of study subjects; running performance was measured for just two young chimpanzees and two people. And the basic premise of the studies, that energy efficiency was the basis for the shift from four-legged to two-legged walking, may also have been fundamentally wrong. According to Leonard and Robertson, we should think about what the earliest humans needed to do with their energy, rather than about energy efficiency per se. If the protohumans were marathon walkers, physiological efficiency would have been critical. But if the earliest stages of becoming bipedal were brought about by some other factor and involved short distances or occasional walking on two legs without full-fledged distance walking, efficiency may have been less important.

In general, running is more energy costly than walking, and a running person should expend calories in direct relation to speed, as happens in quadrupedal animals. But this is not the case in humans. In the range of speeds used by people when running slowly for long distances — about 5 to 8 mph — no speed is optimal, and efficiency does not rise at slower speeds. Why this is true isn't completely clear. One major difference between running and walking lies in the use of the leg tendons, which act as powerful springs during running. The bouncing of a running animal gives back some energy that can be used in the next stride.

The biologist David Carrier of the University of Utah has argued that two-legged running established a new relationship between breathing and striding. A running quadruped tightly coordinates ventilating the lungs and taking strides. The metabolic demand for oxygen is usually accomplished by breathing in lockstep with each running stride. Every time the feet of a racehorse hit the ground, the muscles of the thorax, within which the lungs are ballooning in and out, absorb the shock. This constrains the racehorse to one breath per stride cycle.

Humans have decoupled running speed from breathing. People can take from one to several breaths for each stride or one breath for several strides. At higher speeds bipeds can adjust their breathing rates accordingly. A running person is a multispeed bicycle with seamless flywheel operation. The lack of one optimal running speed may have allowed early bipeds to jog after one sort of game animal while sprinting after another, casually switching gears in ways that a deer or horse cannot. This in turn may have led to further adaptations for meat eating or long-distance foraging or other traits that we consider fundamentally human.

HOW WE WALK

Walking upright is as fundamentally human as any other trait that we possess. But assembling a human body capable of walking is like transforming one jigsaw puzzle into another. It can be done only through many generations of selection for new pieces, selected one at a time until the picture on the puzzle's face slowly becomes an altogether new one. At each small step the puzzle must be fully functional and never a useless mishmash of what it is slowly becoming and what it used to be. First, the move from quadruped to biped required a major shift in the center of gravity. Balanced upright posture requires the center of gravity to float somewhere above the area formed by the feet planted on the ground. In a chimpanzee this center is somewhere in the torso's midsection, suspended between the arms and legs. In people the center has shifted upward and aft from the chimp's, passing just above the bottom two bones in the vertebral column. This simultaneous backward and upward shift was critical: You can imagine the reproductive success of an ape whose kilter is even slightly off when fleeing a predator or fighting over an intended mate.

A well-designed biped should be able to stand at ease and be perfectly balanced. Because our vertical height is dramatically greater than that of any quadrupedal primate's, and because our build is so long and thin, the center of gravity must provide balance as fine as that required to spin a dinner plate on a broomstick. A chimpanzee's center of gravity passes through its gut midsection

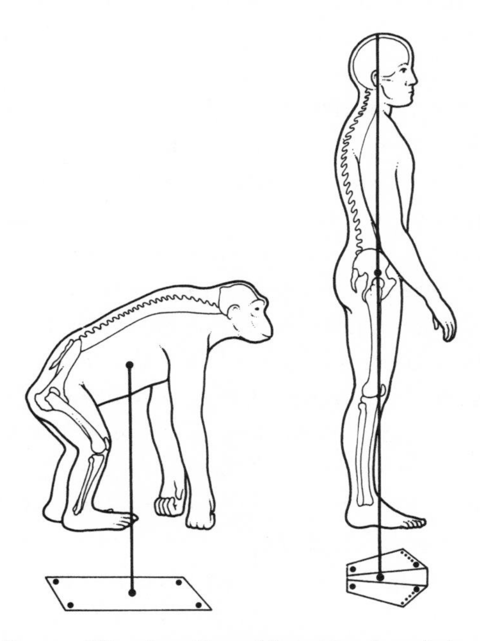

The very different positions of the center of gravity in a chimpanzee (left) and a human.

and up to the sky like the pin that impales a mounted butterfly. The human center of gravity is completely different; it passes vertically through the feet, then the hips, then just in front of the vertebral column and shoulders, through the head in the region of the temples, and up to the sky through the ear region. The result of this new orientation is a biped that can stand in place for hours, exerting only 7 percent more energy standing than lying down. Four-legged animals, by contrast, burn more energy when standing, because their planted legs remain semiflexed, requiring constant muscle action to remain balanced.

ALL IN THE HIPS

The single most dramatic change that our ancestors underwent in becoming human was in the hips. A quick look at an ape pelvis and a human one shows even an untrained eye that something major happened in the transition from them to us. Natural selection took the package of bone and muscle that bears the weight of an ape and twisted it in early humans to create an utterly remarkable system of support for the newly upright walker. The process included warping the shape and form of the ape's pelvis into an entirely new structure and taking an obscure little muscle in the ape's thigh and morphing it into the largest muscle in the human body.

The human pelvis is shaped like a saddle, worn around the waist and designed to support the body at all times. We all know about the gluteal muscles, the lumps of tissue that we sit on, the stuff our buttocks are made of. Three of these muscles are in each upper leg: the gluteus maximus, gluteus medius, and gluteus minimus. In a gorilla or chimpanzee the latter two muscles allow powerful quadrupedal propulsion. They attach to the top of the femur in the upper leg and to the top of the iliac blade of the ape's pelvis, providing the ape with a wide range of motion. Humans no longer need these muscles for propulsion, because extending the hip muscles is not what propels us forward. So this muscle group was freed to evolve into a new alignment and role, that of providing stability while walking upright.

Try this: Walk slowly across the floor and notice that at each step you spend a moment suspended on only one leg, while the other is reaching out for the next step in front of you. The split second during which only one leg is on the floor reminds us how tenuously we stride upon the earth; it is a moment of extreme instability. Your gluteals support you by first pulling your leg away from your body's midline. If not for the realigned gluteals along the sides of your hips, holding you steady as you begin to list to starboard or port, you would be thrown off-kilter as you struggled to maintain balance. Being able to walk is really about being able to stand momentarily on one leg while the other catches up.

When a chimpanzee stands upright, its knees are bent, and, lacking properly positioned gluteal muscles, it must lean forward for balance.

Try walking fifty feet in ultraslow motion. You'll see quickly that standing on one foot without listing a bit is no easy job, and if you touch a hand to your thigh, you'll feel your gluteals tensing to hold you upright.

Chimpanzees fail this test badly. An ape standing upright is forced into a side-to-side rocking movement. Every bad movie ever made about apes, beginning with *King Kong,* has parodied this silly posture. The set of muscles that propels apes forward has been co-opted by evolution to stabilize our human hips. Meanwhile, the remaining gluteal muscle, the gluteus maximus, has migrated in humans. This enormous piece of tissue wraps your rear end like a muscular sarong. Its role is to provide stability, this time to keep the torso upright and smoothly stable while walking.

All these muscle realignments are inextricably linked to radical changes in the body's bony skeleton as well. I remember vividly the first time that I compared a gorilla's pelvis and that of a human. I was sure that I had the wrong bones in my hand. The gorilla's gracefully long and thin main pelvic bone, shaped like a canoe paddle, metamorphosed over a few thousand ape generations into a short, broadly curved human pelvic girdle. Today, the pelvic bones wrap around your waist, acting as a cup in which most of your lower abdominal organs sit. In apes and many other primates the bones extend up the back of the torso, providing extra bony area for the muscles of leg extension to attach. Humans do not need powerful hip extension, and natural selection co-opted the bony features for another purpose.

This purpose is, again, stability and support during walking. Your pelvis is actually made of six bones, matched three on a side, that fused during fetal development. The three — pubis, ischium, and ilium — can still be discerned easily. The pubic bones, which extend from your outer to inner groin, meet at the body's midline with a piece of cartilage connecting them. During childbirth, this cartilage connector loosens enough for a baby's head to pass through the mother's birth canal. The ischium is that part of the pelvis mainly behind and below your thighs; in some primates the area is covered by a rough bit of callused skin that serves as a built-in seat cushion.

The most critically transformed part of the human skeleton that aids in bipedal walking is the ilium. The iliac blade extends around your waist in a broad saddle. Instead of lying more or less flat against the back, as in a chimpanzee, the iliac blade began to flare outward sometime early in the transition from ape to human. In a chimpanzee this long narrow blade aids climbing, because important climbing muscles attach to it. This allows a chimpanzee to pull its weight upward through a tree crown by using its arms as well as its legs. Only in humans has the blade of the ilium become wider than it is high.

By the time early hominids emerged more than four million years ago, the pelvis had already assumed broader, shorter, near-modern proportions. The saddle shape of the newly rotated iliac blade provided space for the attachment of the gluteals, which now ran down the sides and back of the thighs instead of the ape's

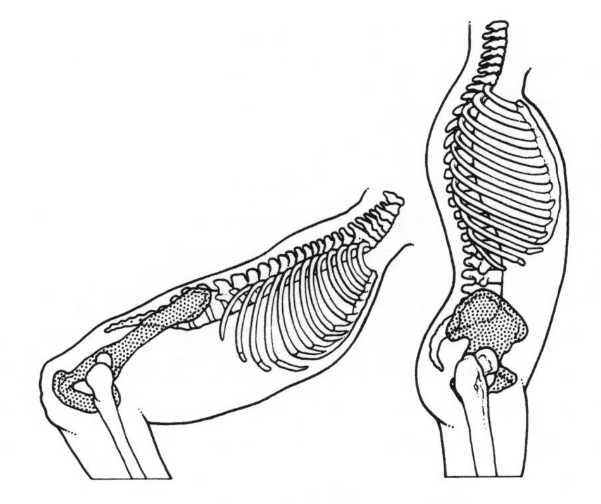

The position of the hips in a chimpanzee (left) and a human.

back-only orientation. At the same time, as the ilium rotated, it became more concave, forming a cone inside the pelvis where your intestines sit. All the machinery of walking was in place.

Other differences between the ape and human skeletal adaptations to walking mainly involve differences in proportion. The human arm bones have shortened relative to an ape's, just as the lower limb bones lengthened. The bones of the fingers and toes lost the pronounced curvature that they have in a tree-climbing ape, flattening into the dexterous fingers and ground-walking toes that we possess. Our fingers have shortened relative to the thumb, so that our fingertips meet easily at one point a few inches out from the wrist, unlike a chimpanzee's elongated digits. And our big toe has migrated from the ape's thumblike opposed position to the same plane as our other toes, providing a better alignment for walking support (still, in societies where people go barefoot, the big toe remains a bit splayed and is used in climbing trees and even holding implements).

TAKE A DEEP BREATH

Natural selection has turned some of the lemons of becoming bipedal into evolutionary lemonade. Some traits may have paved the way for revolutionary changes that were only nascent when we stood up and walked. A human diaphragm is a membranous web attached to the rib cage and vertebral column and located on top of the body's major internal organs. It contracts to help the lungs draw air and life-giving oxygen into the body. This contraction simultaneously helps blood circulate upward toward the heart. The form and function of the diaphragm are much the same in humans as in gorillas. But as we saw earlier, bipeds have a potential advantage over quadrupeds in decoupling their breathing from their locomotion. This decoupling may have freed the human respiratory tract millions of years afterward for a crucial function: the co-option of the adjustable flow of air for speech.

With the rate and depth of breathing no longer confined to the striding pattern of a quadruped, the raw material was in place for other changes to occur. The pharynx — the uppermost section of the esophagus (a food-swallowing tube that opens up at its top end to connect with the nasal cavity and larynx) — became extensively restructured and enlarged in bipeds. The spinal cord also became thicker, allowing even better motor control of breathing. Some observers believe that without these changes in the anatomy of the respiratory and nasal tract, speech would not be possible, much as it is not possible for a great ape. The ape also has other anatomical problems that prevent talking, however, including an ill-formed larynx and palate.

All evolutionary biologists believe that behavioral change precedes and drives anatomical change. Did some early form of language, and the reproductive benefits that its practitioners reaped, drive the changes in the hard anatomy? We don't know, because the soft tissue of the early creatures that might have exhibited language didn't become fossilized. This leaves us speculating about the origins of language. We don't know what the function of early, rudimentary language would have been, or if it was vocal or gestural at the outset. The debate about when language-using hominids arose is unsettled, with the range of opinion between five million and 50,000 years ago. We can be sure only that speech evolution was a wonderfully fortuitous by-product of walking upright.

THE DOWNSIDE OF BEING UPRIGHT

Anyone who thinks that six million years of human evolution has created an optimally designed, anatomically perfect human has never looked closely at his own body. For every benefit that natural selection crafted into the new body plan, trade-offs left the new organism with anatomical or behavioral dilemmas. As the biped gained stability, it lost power. As it gained energy efficiency while striding, it lost that efficiency while climbing. For pregnant women the trade-offs were terrible. Natural selection reshaped the pelvis to accommodate new muscle functions but also contracted the birth canal relative to the size of the infant skull that must squeeze through it. Human and ape experiences with birth differ dramatically because of the shape of the birth canal, which is itself a function of being a biped. As the transition from ape to human pelvic anatomy was happening, the birth canal was squeezed within the shortening, broadening iliac blades.

It may seem natural for a caring man to join his wife in the delivery room for the birth of their child, but a man's presence is a Western aberration in the historical context of childbirth. In the late 1980s I lived in a village in the desert of Rajasthan, India, where my wife, a cultural anthropologist, was doing field research. As she got to know the women of the village and became fluent in their dialect of Hindi, they shared with her their secrets of birthing. Childbearing in this culture, as in thousands of others, is a communal affair for women, one to which men are not invited. When a woman was near term, a village midwife and other women of the community gathered with her to help in the birth of the child. All sorts of inside information, all of it passed orally across the generations, came into play. In the nurturing company of several other relatives and friends, the woman would give birth, aided by the essential knowledge of other past and future mothers.

This level of cooperation is born of necessity as much as goodwill. Women who assist in the birth offer physical, verbal, and moral support and can expect the same intimate assistance when it is their own time to deliver. Without birth attendants the fatality rate to both mothers and neonates would be far higher; breech (feet-first) births and assorted other anomalies would result in certain death to the infant. In the words of the anthropologist Karen Rosenberg of the University of Delaware, such assistance is a cultural solution to a biological problem. Rosenberg and the anthropologist Wenda Trevathan of New Mexico State University documented birthing patterns in nearly three hundred cultures in all corners of the world. In more than 90 percent of them, women gave birth while attended by other women. In the remaining few, experienced mothers sometimes gave birth unassisted, but first-time mothers almost never did. Our stereotypes about village women putting down their garden hoe, having a baby, and returning calmly to work are largely fiction. Everywhere in the world human birth is an ordeal.

Among our great ape relatives, however, birth is definitely not such an ordeal. A female chimpanzee undergoes obvious discomfort in the latter stages of labor. She turns her body this way and that, perhaps to find the least painful position. But the moment of birth is high speed compared with the human experience. The mother chimpanzee reaches down, withdraws the neonate from her birth canal, and draws it up into the cradle of her arms, all in one fell swoop. She bites off the umbilicus, and her new baby joins the world.

What's more, whereas the birth canal of all other primate mothers is an oval from front to back and all the way from the top of the canal to the vagina, the human birth canal is an oval from front to back only partway down. Then it takes on a variety of shapes before exiting the birth canal in circular form. As humans evolved, the birth canal became warped from a longitudinal oval to the variety of shapes it now assumes along its path. As a result, the head of an emerging newborn must rotate as it is pushed down the birth canal; without these twists the baby could not exit its mother.

A baby chimpanzee emerges straight on and face up. This allows the mother to see her baby and to deal with unexpected problems. She can see whether the umbilical cord is wound around her infant's neck and remove it, or wipe mucus from the baby's mouth and nose. And the infant can provide some assistance by reaching up to her with outstretched arms. By contrast, a human baby emerges in a rear-facing position, invisible to and unreachable by the mother. Even if the mother had the wherewithal at that moment to guide her baby out of the birth canal, she would risk inflicting spinal injury to it because of its awkward position. She cannot see the umbilicus or the baby's face.

Assisted births must have begun as soon as the changing pelvic anatomy made birth mechanically difficult, and mothers began to practice reciprocal altruism toward one another. Exactly when rotational birth evolved is not known. If Rosenberg and Trevathan are right, one thing is certain: The earliest hominid females carried out the first medical procedure in the history of humanity.

THROWING A CURVE

The new and awkward alignment of the spine created an additional complication. Like a little suspension bridge, the chain of twenty-four vertebrae that make up the backbone, along with fused sacral bones and coccyx at the lower end, provide the body with supple yet strong support. Between each vertebra is a disk of collagen fibers, shock absorbers that account for one quarter of the length of the backbone.

Natural selection built curves into our backbone — snaking its way from the neck to the hips — that no other mammal needs. A straight up-and-down spinal column doesn't work for a biped whose center of gravity must be kept neatly above the feet while maintaining maximum mobility. A chimpanzee's backbone does not have the four main curves that ours possesses; a chimpanzee doesn't need the curves to maintain its center. Its head is suspended in front of its body; ours must bobble comfortably atop it. A mild forward curve flexes your neck in tune with the movement of your head. At the same time the uppermost vertebrae, those that protect the spinal cord as it enters the skull, serve as a further stabilizer by limiting the range of motion relative to quadrupeds.

A bit farther down the spine a second curve pushes the column back a bit. This is a primary curve, one built into the developing fetus. Another forward curve follows, the so-called lumbar lordosis, followed by a final flip at the tail of the backbone. The lumbar lordosis is more pronounced in women than in men and is the source of many ills. Lower back problems, perhaps the most common chronic ailment in the human species, often originate here. When a woman is pregnant and has a heavy sack of fluid and fetus throwing her center of gravity so far forward that the lumbar curve cannot compensate, lower back problems are rife.

At the same time some researchers believe that the lumbar lordosis, combined with the sharp backward angling of the woman's sacrum and the wedge shape of the last two vertebral disks, allowed nonrotational birth in the emerging human. The lumbar lordosis provides a far stronger and more stable system than a straighter spinal column could. And it seems to have been more recently designed than bipedalism. The primate fossil experts Alan Walker and Pat Shipman of Pennsylvania State University report that while the lordosis of *Homo erectus* more than a million years ago closely resembled our own, the lumbar vertebrae of the earliest hominids were smaller and less well developed than those in more modern people. Strangely, the earliest hominids had six vertebrae in the lower back, rather than the four of an African ape or the five of a modern person. Walker and Shipman speculate that the change from four to six occurred as an offshoot ape evolved into an early human and may have incidentally paved the way for the later development of the lumbar curve. These differences suggest that the gait of earliest humans was probably different from modern walking.

BE COOL

Dogs pant when they're hot, slobbering all over your carpet as they do. We sweat, sometimes profusely. A baseball pitcher can lose ten pounds of body fluid while putting in nine innings on a hot July afternoon. People who lose the ability to sweat — the New York Yankee great Whitey Ford had this medical problem surface late in his career — are in grave danger of lethal overheating. When early humans began to walk upright, their walking no doubt took them out of the forest and into blazing tropical sunshine. Like any other warm-blooded animal, they must have had a way of cooling themselves.

Many mammals use a complex system of transferring heat from the blood of the arteries to the cooler blood of the veins as it returns to the heart. But humans don't have this system. When early humans first began traveling outside the forest, they had to find a way to cool off. A four-legged animal walking long distances in hot equatorial sun exposes its broad back to the sun, a sure recipe for hyperthermia without some adaptation for rapid, effective heat loss.

Two-legged walkers literally are in a better position: The sun strikes only the top of their head and shoulders. Even the slight height difference between the head of a quadruped and that of a biped puts us in significantly cooler air, because wind speed is higher and the temperature is cooler than near the ground. Being naked rather than hairy helps too, because hair traps heat.

But standing tall created a new heating problem — the need to get blood up to the head, counter to the force of gravity. Most animals that constantly shift from horizontal to vertical orientations have evolved features of their circulatory systems that combat the pull of gravity. When snakes climb upward, the pattern of blood flow to the head is radically altered by circulatory adaptations, including a forward-positioned heart that is designed to keep blood flowing. The same thing happens when a giraffe splays its legs, leans its elegant neck down to a waterhole, and then raises it. Special valves, tissue wraps, and pumps prevent blood from pooling in the lower body and boost it up to the hard-to-reach skull. The effect of gravity on blood vessels is identical to the effect of gravity on any column of liquid and is called hydrostatic pressure. It affects the way any fluid, including blood, behaves in a tube that's standing versus one that's lying on its side.

When you lie down, blood drains away from your skull through your neck's jugular veins to the heart. But when you stand up, the blood seeks a different escape route through a vast network of veins surrounding the spinal cord. This network, the vertebral plexus, extends from the skull down to the base of your spine. It diverts blood into the smallest vessels, boosting the blood's flow rate and its capacity to move about the body.

This rerouting is what allows bipeds to efficiently transfer blood, and therefore oxygen, from top to bottom against the pull of gravity. When did humans evolve this adaptation? Close examination of fossils often reveals the indentations of soft tissue, including veins, on rock-hard fossilized bone. Fossils fall into two distinct groups. The more modern parts of the human lineage, from Dart's *Australopithecus africanus* onward to modern *Homo sapiens*, show clear evidence of a vertebral plexus. The other, earlier group shows no such evidence. Instead, earlier hominids possess an enlarged sinus system that shows up as a deep groove inside the rear portion of the brain case. This groove is rarely found in modern people, or in four-legged primates, suggesting that it was the method that natural selection chose for draining blood from the skull of the earliest bipeds.

The anthropologists Dean Falk of Florida State University and Glenn Conroy of Washington University applied this knowledge to the human fossil record. They reasoned that the routing of the circulatory system provides keys to the lives and habitats of the earliest humans. When the shift to bipedalism happened, it must have been accompanied by a change in the way that blood was moved up and down the now-vertical column of the body.

Brains in particular need to be kept cool; overheating poses a greater risk to the brain than to other parts of the body. Falk suggested that the vertebral plexus evolved to cool down a rapidly expanding brain as emerging humans moved out onto open, sun-soaked grassland. The idea is provocative: the circulatory system as a radiator designed to keep a growing brain cool. This enabled increasing brain expansion in one lineage but not in another. If Falk is right, there were two ways to drain blood from the human skull in prehistory. One option was taken by the early hominids that were direct forerunners of modern people. Another was used by other lineages that went extinct without leaving modern descendants. Thus we can use the vertebral plexus as a test of which species of our ancestors were our direct forebears and which were left on the cutting room floor of evolutionary history.

4 THE EXTENDED FAMILY

Progress, therefore, is not an accident but a necessity.
— Herbert Spencer, *Social Statics*

IN 2000 A TEAM of fossil researchers led by Meave Leakey of the National Museums of Kenya announced yet another trophy for the Leakey family mantelpiece. Since 1998 the scientists had unearthed new fossils at a site called Lomekwi, on the western shore of Lake Turkana. The creatures had lived smack in the middle of the reign of the famed fossil human Lucy and her brethren. The bones of the new fossil revealed an early human with a face and teeth that resembled an early representative of our own genus, *Homo*. Leakey deemed the new fossil sufficiently different from those previously discovered that she bestowed on it not only a new species name but a whole new genus — *Kenyanthropus platyops* (the flat-faced man from Kenya). The Leakeys crowed about their find: With its mosaic of chimpanzee and human features, *Kenyanthropus*, not Lucy, might be the forerunner of us all, they claimed. And nothing could have been sweeter than finding *Kenyanthropus*, given the long-standing rivalry between the Leakey family and Donald Johanson, who had found Lucy twenty years earlier. Scientists are now debating whether the Leakeys' claim is a bit of taxonomic overreaching: *Kenyanthropus* may be just a very slight variant of Lucy's species. But clearly the place was getting crowded about four million years ago, and that fact points to one thing: Just as gorillas, chimpanzees, and bonobos all live in the same region of Africa today, many kinds of early hominids lived there contemporaneously too. Those specimens that are known, plus those yet undiscovered, formed a golden age of the hominids.

That's why the old-fashioned notion of a ladder of evolution is so wrong. We didn't become human by climbing rungs; we're here because we avoided the shears when natural selection trimmed our rapidly growing tree of lineages. No single missing link between apes and humans exists. The human species was assembled from a vast gene pool that was distributed over a wide geographic area. A new trait — slightly upright posture, for example — might arise in one population because of a mutation and would spread as individuals in that group migrated to other groups. The resulting gene pool would become a constantly shifting code for what the animals looked like. No single member of the species could be called the progenitor; traits accumulated to create a mosaic body plan.

We never see the earliest member of a new lineage; the fossils we find typically represent their descendants many generations later. Indeed, a team of biologists has recently proposed that the earliest primates emerged tens of millions of years earlier than we had thought, an error that occurred because we tend to forget how fragmentary the fossil record is and how likely it is that many early species remain undiscovered.

We are driven to assemble order out of chaos. If I give you a box overflowing with items of all shapes and sizes, your first instinct will be to sort the contents by color, size, uses, and whatever else seems appropriate. We are categorizers by nature. We need to pigeonhole things. We're also linear thinkers, looking for logical connections between points A and B, whether they exist or not. Even the best of us dismiss or ignore the multiple levels of a problem in order to find a simple, direct, logical path.

But evolution by natural selection tolerates, even favors, tremendous variation; rather than winnowing, it tends to foster diversity. The tree forked repeatedly, and in the end the one remaining twig was that of *Homo sapiens.* Hominids flourished only because they emerged at the right place at the right time. As with every other evolutionary success story, of which we are the latest but probably not the last, context was everything.

There was no direct linear progression from quadruped to semibiped to fairly efficient biped to full-fledged biped. Many new models of two-legged walkers failed, and puzzled paleontologists find the fossils of others that succeeded for a time. A few succeeded grandly, occupying a variety of habitats in a large area. The early hominids that we know about are *Australopithecus afarensis* (Lucy and kin), *A. anamensis* (another of the Leakeys' recent discoveries), the recently discovered *A. garhi* (apparently the first meat eater), the robusts (named for their oversized molars), *Homo habilis* (the earliest member of our own genus), the enigmatic *Ardipithecus ramidus* (a prime candidate for the most primitive hominid), and now perhaps the Leakeys' brand-new *Kenyanthropus platyops* specimen. In addition, we have found several fossils that are of ambiguous ape or human identity: *Orrorin tugenensis* and "Toumai," *Sahelanthropus tchadensis.*

This is a modest array, only several species, each with its own set of adaptations to its time and place. The diversification of humankind was small by mammalian standards, so understanding bipedalism is difficult; our ancestors comprised a relatively small number of species. But Earth's history does feature a wildly successful proliferation of bipeds. So before we consider the evolutionary significance of the earliest humans, let's look at the only other diversification of large-bodied bipeds in Earth's history.

LESSONS FROM THE JURASSIC

Dinosaurs, as we all know, came in all shapes and sizes, from roadrunner-like two-legged walkers to lumbering four-legged leviathans. The first dinosaurs were all small, carnivorous, and bipedal. Their forelimbs withered over eons to become grasping hooks that probably aided them in catching and eating prey. All the gargantuan forms that every seven-year-old knows by heart evolved from these ferocious little upright runners.

Dinosaurs are the only diverse group of animals other than hominids that evolved bipedalism as a strategy for making a living. Among the bipedal dinosaurs were massive predators such as *T. rex* and *Gigantosaurus* (which recently dethroned *T. rex* as the greatest predator ever), lumbering bipedal herbivores such as *Hadrosaurs* (the duck-billed dinosaurs) and *Parasaurolophus* (the dinosaur wearing the tall bishop's-miter crest), and a host of small- and medium-sized carnivorous bipeds, including raptors. The quadrupedal forms evolved from these bipeds. And some were sometimes quadrupedal but usually bipedal, such as *Iguanodon.*

The success of bipedal dinosaurs was partly historical accident. Natural selection worked with an early two-legged form and created modifications in body size and teeth, as well as all manner of other alterations in body parts. Many variations on the theme of upright walking evolved during their reign. Some bipedal dinos stood erect while others were more horizontal, and some ran as swiftly as a racehorse while others moved more like hippos. The duckbills were almost certainly herd-living plodders that may have spent part of their day on all fours. They were among the few bipedal dinosaurs that were herbivores, not predators. Paleontologists debate whether *T. rex* lumbered along, scavenging huge carcasses of animals that it was too slow to capture alive or whether *T. rex* was an efficient predator that pursued its prey the way a tiger runs down a deer.

In late 2000 a team of paleontologists, led by David Berman of the Carnegie Museum of Natural History, announced the discovery of a 290-million-year-old fossil reptile. The creature, which the team named *Eudibamus cursoris* (the dawn runner), had clearly been a swift runner. Its hind limbs were much shorter than its forelimbs, and they were arranged so that, like any good runner, its legs straightened when fully extended. *Eudibamus* probably reared up on its legs and sprinted when it needed to, resting or walking on all fours at other times.

When running, the little reptile maintained a nearly vertical posture that did not appear in the precursors of the dinosaurs until sixty million years later. So eons before the first dinosaur emerged, bipedal reptiles were running in the vertical posture that dinosaur lineages took millions of years to evolve. This tells us that becoming a biped is not necessarily a long and slow progression from mediocre four-leggedness to elegant uprightness. Rather, it's just what natural selection did with the material on hand, a successful experiment among many.

When we look at dinosaurs, we don't expect to see progress from less bipedal to more bipedal forms. Nor do we expect a uniform type of two-legged walking among all species or in all eras. We understand that different kinds of upright walking suited different climates, habitats, and diets. The myth of bipedal progress, so entrenched in the study of human posture, is absent from dinosaur research. We view dinosaurs as evolutionary failures and ourselves as evolutionary success stories because dinosaurs are extinct. But dinosaurs flourished for 150 million years, about thirty times longer than our own human lineage has been on Earth. We view early human bipedalism as evolving *toward* the modern human form because we are such good bipeds. Our view of the evolution of bipedalism is utterly and misleadingly humancentric. Dinosaurs are a healthy corrective to our deeply held view that bipeds are better than quadrupeds and that full-fledged bipeds are better than partial bipeds.

THE BIPED ZOO

The period of Earth's history from five million years ago until just under two million years ago was the heyday of human bipedal biodiversity. About six million years ago the ape-human fork in the road appeared, with a new lineage whose primary adaptive shift was the adoption of upright posture. A whole array of two-legged forms then fanned out across the landscape of East Africa and perhaps to the south and west as well. Fossil discoveries in the

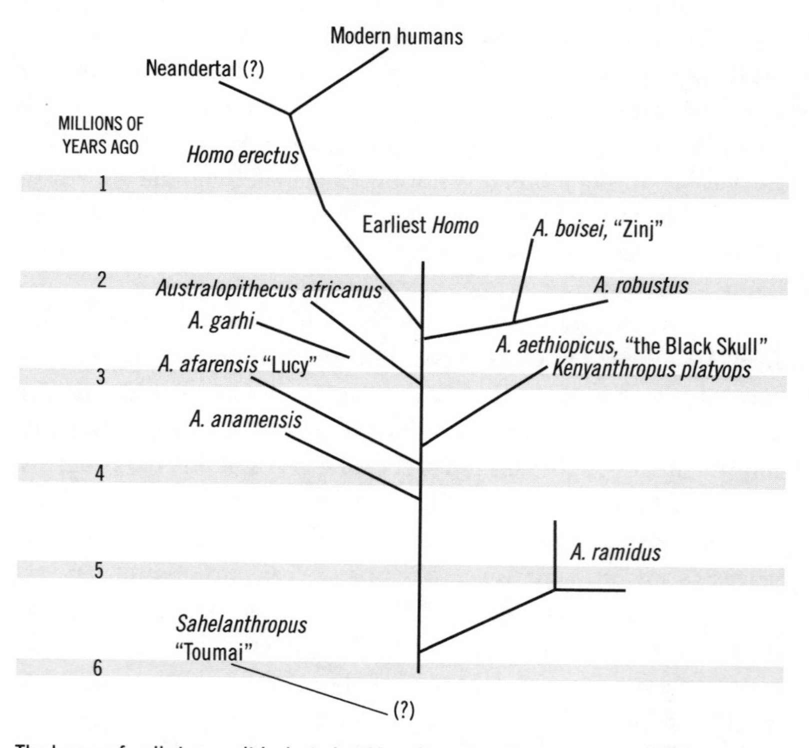

The human family tree as it looks today. Many branches have uncertain connections to the trunk and to each other. Also note the uncertain status of Toumai, the fossil ape (or protohominid) from Chad.

past few years indicate that bipedal walking occurred very early in our evolution.

In 1994, long before the Leakeys had claimed their *Kenyanthropus* to be the rightful heir to Lucy's throne, they found a different fossil human. At two sites in northern Kenya, Kanapoi and Alia Bay, a team directed by Meave Leakey found fragmentary remains of a primitive hominid. It looked like Lucy in its lower limbs but was more chimplike in its teeth and jaw. The Leakey team named the specimen *Australopithecus anamensis* (southern ape man of the lake, in the local Turkana language) and later dated the fossil to about 4.1 million years ago. This made it just a bit older than Lucy, the oldest biped known at that time. Although there was some debate about the find, most authorities recognized the new species in the mid-1990s as the earliest certain hominid.

You might reasonably ask whether we can be sure that two very similar species in the fossil record were not really just one. After all, while chimpanzees and bonobos differ in aspects of their social life and reproductive physiology, a paleontologist would have a hard time identifying them as different species based solely on their skeletons. And an assemblage of bones containing one large and one small human could be two species or just big males and small females of the same species.

The difference can be resolved with statistics, so long as we assume that the range of sizes of extinct apes and humans is similar to what we see among living ones. But this is not always a safe assumption because of the small number of living species that are available as reference points.

The biological anthropologist Michael Plavcan of the University of Arkansas showed that even the most widely used statistical measure cannot determine whether a jumble of bones should be sorted as one species with wide size variation or as multiple species. Plavcan applied the measure to several well-known species of African monkeys and got the startling result that the test could not reliably sort any of the species. This cautions us that unrecognized fossil hominid species are hiding among the many specimens in the museum cabinets of the world, not to mention in the arid regions of Africa. Hidden within the current evolutionary tree of eight to ten hominid species are, no doubt, more that cannot be distinguished based on the tiny sample of bone fragments that we have. For example, what we call *afarensis* may have been one species in Ethiopia, another in Tanzania, and a third in Chad.

Also, an awkward difference exists between what we call a species in the fossil record and what a biologist would call a living species, so knowing how many species of early hominid existed is problematic. We will probably never know the number for certain, because we find human fossils much closer to the surface of the earth than dinosaur fossils, since emerging humans were around for only two million years. Hominids also occurred on only one continent, and only certain habitats are amenable for preservation of fossils (remember that soft tissue and behavior are what make all the difference).

The paleoanthropologist Robert Foley of the University of Cambridge has estimated that the level of early hominid diversity was significantly greater than we recognize, based on the known distribution of early hominids in Africa plus the regions where they probably lived but that have not yet yielded fossils. Until some eager young fossil hunter finds a treasure from areas still untapped, we can only speculate.

We do, however, have a far more informed view of the fossil record and of bipedalism than we did in the 1970s because of the steady accumulation of fossil discoveries. Today we know that by four million years ago hominids were already accomplished bipeds. The frustrating problem is that these fossils shed little light on how and why bipedalism evolved. Therefore, we seek critical information from even earlier humanlike fossils. *Ardipithecus ra-*

midus holds an enigmatic place in the human family tree. In 1994, while working in Ethiopian badlands not far from where Donald Johanson had found Lucy twenty years earlier, Tim White of the University of California, Berkeley; Gen Suwa of Tokyo University; and Berhane Asfaw of the Ethiopian National Museum found a treasure trove — a deposit of fossils belonging to a very early human ancestor. White, Suwa, and Asfaw named their new find *Ardipithecus* (earth-ape, for its ground-living anatomy) *ramidus*. The fossil's age made the find astounding: nearly 4.5 million years old, the most ancient member of the human family yet discovered. More recent fieldwork has turned up more bone fragments that are even older, pushing the dates of the creature back to nearly six million years ago.

Aramis, where the species was found, has turned out to be a fossil hunter's dream, yielding ever more remains as the work continues. Between four and six million years ago Aramis was a dense upland forest inhabited by many animal species that we associate with a tropical forest in Africa today. White, Suwa, and Asfaw's *A. ramidus* made its living as a forest ape does, perhaps as an upright incarnation of a modern chimpanzee or bonobo. We still don't know some of the key features of *Ardipithecus*, because the most complete specimen of the ninety-odd represented in the deposit is still being painstakingly excavated from the delicate rock matrix in which it was discovered. The paper by White and colleagues announcing the discovery described a very apelike hominid, based on its teeth, perhaps with a primitive variation on bipedal posture. Or perhaps we should say it was a hominid-like ape, because bipedalism has long been the benchmark for conferring hominid status on a fossil. White considers *A. ramidus* a potential ancestor of later hominids. When the excavation is complete, the world may gain a whole new perspective of the human family tree. If *Ardipithecus ramidus* is not quite bipedal, the species may turn out to hold the answers to all sorts of questions. All manner of fully or semi-bipedal creatures may have radiated from an ape ancestor, although only one lineage resulted in us.

Another recent fossil discovery announced in 2002 could shake the branches on which sit *anamensis*, *afarensis*, and *ramidus* in the human family tree. A team led by the French paleoanthropologist Michel Brunet found the remains of a primitive hominid, or perhaps ape, in the Saharan sands of Chad, in west-central Africa. The team dubbed the find Toumai (they bestowed on it the scientific name *Sahelanthropus tchadensis*). The date of the fossil is the key to the attention that it generated. A comparison with other sites containing similar fossilized animals yielded an estimate of six to seven million years old. Because the base of the skull does not clearly reveal whether the spinal cord passed vertically through it, scientists don't know yet whether Toumai was a biped or perhaps nothing more than a fossil gorilla. Other authorities think that the ancient date is erroneous and that Toumai is just a contemporary of other, already known, hominids. But some experts think the skull is the most exciting fossil discovery in decades.

In 2000 Martin Pickford of the National Museums of Kenya and Brigitte Senut of the University of Paris announced that they had found remains of a six-million-year-old humanlike fossil in the Tugen Hills of Kenya. The researchers dubbed the fossil *Orrorin tugenensis*, the Millennium man, and suggested that it might be the very earliest bipedal hominid. But most of the anthropological world disagrees with this assessment because of the very apelike appearance of the bones. Although the debate continues, Millennium man may well turn out to be a millennial ape, a long-lost ancestor of the modern chimpanzee or gorilla.

If early hominids need not have been bipeds, early bipedal primates were not necessarily hominids, either. A good example of how standing upright need not be a precursor to being human comes from the fossil ape *Oreopithecus*, the so-called Cookie Monster. The paleontologists Meike Köhler and Salvador Moyà-Solà of the Barcelona University Institute of Paleontology studied the remains of one species of *Oreopithecus*. It lived from seven to nine million years ago on an island in the Mediterranean Sea. Their reconstruction of its foot no doubt made them stop and take a deep breath. The ape's feet resemble those of no ape or human living or extinct but something in-between. The toes stick out to the side the way that gorillas' toes do when they are walking on the ground. This, combined with a widely diverging big toe, makes the creature's foot resemble a tripod, designed for standing support rather than the usual ape emphasis on climbing agility. The researchers point to a few other oddities in the skeleton as clues to what the animal was like in life. Recall from chapter 3 that modern humans have wedge-shaped lower vertebral disks. *Oreopithecus* has the same wedging. And its pelvic anatomy resembled that of early hominids more than that of any ape, living or extinct.

Köhler and Moyà-Solà have decided that the creature was a bipedal ape. But they do not believe *Oreopithecus* to be the long-awaited missing link between the fossil apes and the fossil hominids. Instead, they believe that the creature is evidence that natural selection experimented with bipedalism even before the emergence of hominids. Köhler and Moyà-Solà conclude this because of the fossil's age, its unique habitat, and its otherwise distinctive ape anatomy. The excavation site is in what today is Italy. The authors argue that the ape was an ancient example of what happened to many modern creatures that evolved in isolation on island settings. Köhler and Moyà-Solà argue that, lacking predators (none has been uncovered at the site), *Oreopithecus* evolved bipedalism as an easy way to leave the trees and amble somewhat awkwardly about on the ground.

Some researchers have questioned this interpretation of the *Oreopithecus* remains, arguing that *Oreopithecus* is just a good example of how a vertical climbing ancestry produces a skeleton that superficially resembles a biped. But if Köhler and Moyà-Solà's reconstruction is correct, it makes perfect sense. Bipedalism is a strategy for foraging that works in the right environment. Nothing is special or sacred about bipedalism relative to four-legged posture and gait. The story also suggests that early in hominid evolution, bipedalism may have taken many forms in many habitats, each suited to a type of feeding or traveling behavior, similar to the way different tree-living animals today have a wide variety of adaptations to climbing.

DEVOLVING?

While most of the paleoanthropological world accepts *A. afarensis* as a candidate for the trunk of the hominid family, recent dissenters have come forward to argue otherwise and have often fallen into the magical thinking that seems to define how we view the human fossil record. When the South African researcher Lee Berger described a primitive leg bone from *Australopithecus africanus*, he concluded that *africanus*, although a more recently emerged species than *afarensis*, was more primitive in its form of bipedalism. This presented a conundrum. How could a species with a more apelike walking posture have arisen *after* a more modern form had? Some experts believe that Berger simply did a poor analysis of the specimen — a tibia from Sterkfontein on the South African plains. Berger himself believes in the possibility that his specimens underwent "an evolutionary reversal." Such thinking is driven by a simplistic understanding of how natural selection works and perhaps also by a desire to make South Africa, rather than East Africa, the cradle of humanity. The idea that evolution worked in reverse in South Africa is not inconceivable, but it's extremely unlikely in the face of better explanations. Natural selection was tinkering with posture and gait 3.5 million years ago; in southern Africa it produced one form and a bit later in eastern

Africa it produced another. The imperative that fossil hunters feel, to line the species up like fire fighters climbing a ladder to a burning building, is a reflection of their own worldviews and the source of much confusion.

A ROBUST WORLDVIEW

I will discuss Lucy and her kin, *A. afarensis*, in detail in the next chapter; her species enjoyed life on Earth for one million years, from nearly four to just under three million years ago. This is six times longer than *Homo sapiens* has been on Earth. The discovery of a new australopithecine, *A. garhi*, that lived 2.5 million years ago in East Africa, complicated the debate about *afarensis*'s immediate descendants.

The Ethiopian fossil expert Berhane Asfaw and his former mentor Tim White found the fossil, which was announced in 1999. They found the remains very near White's renowned *Ardipithecus ramidus* location in the badlands of Ethiopia. *A. garhi* is quite different from any other known hominid fossil. It possesses a projecting face and very large front and back teeth. Its arms were long, almost apelike, but it was also leggy, more so than other primitive humans. The new species is a potential descendant of *afarensis* and a good candidate as an immediate ancestor of the early members of the genus *Homo*. *A. garhi* is also significant for what was discovered with it: stone tools. These simple stone tools, probably made and used by the chimpanzee-like creature, push back the earliest known date of tool use even further than previous finds and tell us that even very primitive humans were butchering and eating animal carcasses. On the whole, it throws the human lineage into great confusion, because no one expected to find a new hominid in that region or that time period. As such, it is a powerful statement about how nonlinear and incomplete the fossil record is.

Immediately following the time of *Australopithecus garhi*, the human family tree branched. The road less taken, as it were, led to a unique group of humans, collectively known as the robust australopithecines, so named for their huge and powerful molars.

The first species discovered was also the career-making find by Louis Leakey and Mary Leakey. Louis Leakey, the son of English missionary immigrants to East Africa, had grown up in Africa and long believed that the key to human origins would be found there. He toiled for years in a variety of places on a shoestring budget, finding little to keep his quest alive. Beginning in the 1930s, he and Mary, his wife, made an annual trek in an old vehicle to Olduvai Gorge, which was accessible in those days only during the dry season. During these expeditions of several weeks, the Leakeys worked long hours in obscurity, believing that eventually their efforts would be rewarded.

In 1959 Mary found a remarkable skull. While the Leakeys might have expected a chimplike skull to characterize their first early human find, the skull she found was unlike that of any ape they had ever imagined. The skullcap displayed flaring cheekbones, tiny front teeth, and a pushed-in, dishlike face. A broken ridge of bone ran lengthwise along the top of the skull, giving the creature's head the look of a centurion helmet in the making. This crest expanded the surface area of the skull so that additional space was available for chewing muscles. If you touch your finger to your temple and make a chewing motion, you will feel the fan-shaped temporalis muscle at work beneath. The addition of the bony crest allowed the creature, *A. boise*, to possess the powerful chewing muscles that it needed to crunch hard-shelled objects eaten by other animals inhabiting its realm, perhaps including other early human species.

The other striking trait of the new fossil was the tiny size of the front teeth, which were little more than stubs, in stark contrast to the enormity of the molars. The small teeth led the Leakeys to call the specimen Nutcracker man, or *Zinjanthropus boisei.* Zinj, as the scientific community called the skull until it was recognized as a member of the genus *Australopithecus,* created an enormous stir in the anthropological world. For the first time solid evidence finally existed of early humans' origin in East Africa. The animal's location further validated the South African discoveries decades earlier.

Zinj also showed that early humans were a diverse lot, its robust skull contrasting with the slender delicate features of Taung. And the Leakeys' find demonstrated that early humans had ranged all across Africa. *A. boisei* had a counterpart in southern Africa, *A. robustus,* discovered in the 1930s in a limestone quarry by the physician-fossil hunter Robert Broom.

While *afarensis, garhi,* and Dart's *A. africanus* probably ate plants and some meat (the evidence for meat eating is strongest for *garhi* and circumstantial for the other two), the robusts' molars show the wear and tear of a tougher diet. Examination by high-power microscope shows a pattern that is distinctive of eating extremely fibrous and hard-shelled foods, which may have enabled the robusts to avoid competition with other early humans. The robusts arose at least 2.5 million years ago and were still living in southern Africa a little more than one million years ago. In East Africa *boisei* was living in the same place and time as early members of our own genus, *Homo* as late as the reign of *Homo erectus.*

Both these robust species may have had a common ancestor in East Africa named *Australopithecus aethiopicus.* Discovered on the western shore of Lake Turkana in Kenya in 1986 and nicknamed the Black Skull because of the mineralized patina on the first specimen, *A. aethiopicus* bears an uncanny resemblance to what the ancestor of the robusts and the descendant of *afarensis* should look like. This creature may be the bridge between the robust lineage and the rest of human genealogy. It possessed a longitudinal crest, flaring cheekbones, and evidence of large molars, as well as features reminiscent of earlier hominids. We assume that the two more recent species, *boisei* and *robustus*, are the direct descendants of *aethiopicus*, but as Melanie McCollum of Case Western Reserve University has shown, this may not be so. McCollum used a detailed analysis of the face of the robusts to show that the southern and eastern species were just as likely to have separate ancestries as the same one. This is compelling, because it again points us away from the simplistic linear evolution model and toward a more realistic picture, a knot braided with the threads of human history.

The trio of robust hominid species thrived for more than a million years of human history, many times longer than *Homo sapiens* has been around. The robusts flourished for a time but faded into evolutionary anonymity. They represent one of the two presumed major forks of our family tree. The other fork led to our own genus and the earliest members of *Homo*. In 1960 Louis Leakey's son Jonathan uncovered the remains of a protohuman so primitive that it seemed to be an immediate descendant of the australopithecines. The fossil's estimated brain volume, about a third larger than that of any australopithecine, led Louis Leakey not only to name a new species, *Homo habilis*, but to redefine the genus *Homo* to accommodate it.

Louis Leakey knew that early *Homo* shared its world with his *A. boisei*, because he found both in Olduvai Gorge in material dating from about the same time. The difference between an australopithecine and an early species of our own genus is subjective and to some extent semantic. They share primitive features that link them to the apes and ourselves. Leakey discovered two dramatic differences between the two. Whereas *boisei*'s strength was in his molar teeth, Leakey's *Homo habilis* had brain power. Not a lot of brain power, perhaps, but with a brain about a third larger than that of any previous human, *H. habilis* elevated itself above the ranks of glorified apes and into a more human way of life. This new way involved technology: the making of stone tools. Although we now know that *A. garhi* probably was also a maker of tools, the earliest members of our own genus were the first that seemed to rely on such artifacts to help them change their environment, by butchering carcasses for consumption. *H. habilis* and several close relatives were still quite apelike, but they had taken a small step toward humanity that may have given them a crucial advantage in competition with other hominids of their era.

Consider human evolution not from our own modern perspective, which is biased by the knowledge of how it all turned out, but from that of a robust hominid who was a historian. She might look at the current state of human development about two million years ago and foresee that the gracile lineage, the one that we know leads to ourselves, would ultimately die out, leaving no trace. Such a historian would see a plethora of ape species, many of them on the decline toward extinction. She would not see a simple branching of human lines; on the contrary, she knows that all manner of experiments have come and gone. She would know that the robusts are the pinnacle of human evolution. They had specializations of the body that allowed them to invade a whole ecological niche that the apes and other humans had not. They were flourishing in both eastern and southern Africa and perhaps other regions as well. They coexisted with other human species that had larger brains. But the most successful lineages in the history of life on Earth were notably pea-brained, suggesting that large brain size might even be a detriment to the long-term success of an evolutionary line.

From the viewpoint of a robust australopithecine, the past and the future seemed rosy and bright. And within a few thousand generations they all would have vanished from the earth. So instead, we should turn to the lineage that may have produced descendants whose progeny are alive today.

5
EVERYBODY LOVES LUCY

THE ANNUAL MEETING of the American Association of Physical Anthropologists convened in Denver in the spring of 1994 for a four-day succession of academic slide presentations delivered in darkened rooms by both veteran scholars and nervous students of human evolution. As always, the subplots were many — academic gossiping in the corridors about the latest discoveries, tenure decisions, and hirings. Like many others, I remember only one paper from that meeting: a report of the discovery of new specimens of *Australopithecus afarensis* in Ethiopia. Donald Johanson, Lucy's discoverer and at the time director of the Institute for Human Origins in Berkeley, California, was to describe his team's latest findings on center court. The audience overflowed the auditorium; I sat on the floor in an aisle, trying to see the action over hundreds of heads. The schedule for the afternoon had run late, and at the time set for Johanson's talk, a nervous student, whose lecture was scheduled before Johanson's, had an unexpected standing-room-only crowd seeking to squeeze in for the main event.

Johanson took the podium and described the discovery of the long-sought almost-complete skull of an *A. afarensis*, an early member of Lucy's species. This was a key find, because Lucy and the hominids from Hadar discovered in the 1970s lacked complete skulls (the reconstruction of one skull included estimates of some skull features; the final product had met with criticism in some paleoanthropological circles). The conference paper also gave Johanson a chance to discuss other aspects of *afarensis*, as well as to address some of the more contentious aspects of the debate about human origins. At one point, describing *afarensis*'s lower anatomy, he referred to the acrimonious debate about how Lucy walked.

A research team led by Randall Susman of the State University of New York at Stony Brook had been saying for years that Lucy and her cousins did not walk in a modern upright posture. Instead, the Susman team argued, Lucy's kin had retained the ability and propensity to climb trees for a living in addition to walking awkwardly on the ground. In its most recent study the team had reconstructed Lucy's walking stride. Lucy's foot was almost a third longer relative to her legs than the feet of later hominids, the Susman team noted. To understand this effect on walking, the researchers from Stony Brook studied people walking in specially fitted oversized footwear and found that they moved inefficiently and very differently than modern people normally do.

Now Johanson threw his knockout punch. He alluded to the new Stony Brook research, referring to it as the "noted clown-shoes hypothesis." The audience gasped, then exploded into laughter, deriding both the Stony Brook theory and Johanson's sarcastic lampooning of it.

Lucy is that rare case of a creature who, like an old Hollywood celebrity, becomes an icon even to people who aren't exactly sure who she was or why she was important. In life she was about three feet tall, an apelike creature that lived and died like all the other less-renowned members of her species, *Australopithecus afarensis*. The dramatic 1974 discovery of A.L. 288-1, as she was catalogued, has been described many times. The account usually focuses on Johanson's extraordinary fossil-finding success in the

Afar badlands of Hadar, Ethiopia, or on Lucy's being nicknamed after the Beatles' song played in celebration that night in the field camp, or on Johanson's ensuing discovery of the "first family," a whole group of *A. afarensis* that may have perished together in a flood or other calamity.

The public has paid far less attention to the underlying debate about the meaning of Lucy's remains and those of her relatives. The real story of Lucy has been chronicled mainly in hundreds of data-filled scientific articles in austere publications such as the *Journal of Human Evolution* and the *American Journal of Physical Anthropology*. If I reached into my bookcase and pulled down a handful of issues from these publications, the articles would form an academic version of a shouting match. Veteran scholars, their loyal student disciples, and brash new Ph.Ds eager to make their mark have produced hundreds of contentious articles and conference abstracts.

Although more recently discovered early humans may be our closer relatives, the amount of research done on A. L. 288-1 and the completeness of her skeleton make her the most important early hominid fossil ever found. Everybody loves Lucy; of all the females who have ever lived, none has had so many men — and women — fighting over her so long after her death. She is the Rosetta stone through which anthropologists look for the truth about our origins and the chronological benchmark that we use to gauge all fossil humans' degree of modernity. Nevertheless, after decades and all the scientific studies of her frame, no solid consensus exists about her exact place in human origins. The scientific teams that have warred over Lucy since her discovery are starkly different in their perspectives of her.

Lucy was first presented to the scientific world in the mid-1970s by Johanson and Tim White. She is just one specimen representing *A. afarensis* that lived for more than a million years early in human history. "She" seems to be accurate; Robert Tague of

Louisiana State University and C. Owen Lovejoy of Kent State University recently defended Lucy's femininity after the University of Zürich researchers Martin Haüsler and Peter Schmid argued that Lucy was really male, claiming that she was a small species of human that coexisted with another, larger species. They based their argument on a comparison of Lucy's pelvis with those of other australopithecines and of modern humans. But Tague and Lovejoy showed convincingly that she is unabashedly a girl, with a birth canal.

Johanson and White published their formal description of the famed fossil in 1978, and their further analysis of Lucy's place in our family tree appeared the following year. Owen Lovejoy, along with Bruce Latimer of the Cleveland Museum of Natural History and William Kimbel of Arizona State University, also undertook extensive analyses of the skeleton. Their consensus was that Lucy was a full-time habitual biped.

In another analysis Lovejoy went even further. He concluded that Lucy was adapted only for bipedal walking and lacked adaptations for any other mode of travel. If she climbed at all, she did so in the way that you or I can. Lovejoy even claimed that *afarensis* was as well adapted to bipedal walking as modern people are. He based this claim in part on the *afarensis*'s new alignment of the gluteal muscles. The gluteus maximus had ballooned in size and had shifted above the hips. This restricted Lucy's ability to climb vertically, Lovejoy decided. What's more, he concluded, the changed relationship between the length and tension of the gluteus maximus and other walking muscles would have exhausted Lucy if she tried to use them for anything other than their intended bipedal purpose. Even modest amounts of apelike climbing in trees would have fatigued her. And Lucy did not have the grasping feet of an ape.

Then, in the early 1980s, Susman and his Stony Brook colleagues, Jack Stern and William Jungers, entered the fray. They made measurements of the fossil, using cast replicas that Johanson had made. This evidence consisted of similarities between the length and proportions of the upper and lower arm bones, as well as the alignment and form of the knobs and bumps where climbing muscles attach to the bones.

Stern and his colleagues agreed with the initial assessment of A.L. 288-1: She was a biped who clearly showed some major anatomical adaptations to life on the ground. The Stony Brook team addressed two fundamental questions. First, how accomplished a biped was *A. afarensis?* Did hominids in this group walk fully upright as we do, as Lovejoy believed, or did they represent some intermediate form of bipedalism? Second, did Lucy live entirely on the ground, or did she spend part of her life climbing about in trees? The possibility that Lucy and company were not fully bipedal implied that their lifestyle was not fully terrestrial, either. The nature of our early ancestry hung in the balance. Lucy's skeleton was a gold mine, and the nuggets were the data points that could be elucidated by careful examination of her pelvis and lower limbs, all of which were in a remarkable state of preservation. Anatomical comparisons of Lucy and the other Hadar *afarensis* with *Australopithecus africanus, boisei, robustus,* and the earliest fossils of the genus *Homo* began to fill the pages of major journals.

Stern, Susman, and Jungers sought answers to these questions. They launched their attack with a lengthy paper published in the *American Journal of Physical Anthropology* in 1983. They presented evidence that *Australopithecus afarensis* was not a full biped in the modern human sense and that the creature's anatomy showed clearly that it was still well adapted for life in the trees. Lucy, the authors stated boldly, is as close to the missing link as we are likely to find; any member of our lineage that is more ancient is likely to be just an ape.

Stern and Susman did not argue that *afarensis* lived in the trees; they said that she was primitive enough to have the anatomical ability to be at home both in trees and on the ground. They pointed out that the finger bones and some of the wrist bones of the hand were very similar to those of a chimpanzee, and the hand-wrist combination was apelike. A peasized pebble of a bone in the wrist, the pisiform, was elongated in the manner of an ape's; Stern and Susman suggested that this gave Lucy the ability to flex her wrists more than a biped could have and indicated a tree-climbing habit. They also claimed that the metacarpals — the long bones that are splayed in the back of your hand — were those of a tree climber. The picture they painted was of a hand that was used to hang by.

Then there were the hips. Stern and Susman noted that Lucy had a long set of hamstrings, the divided set of muscles that in people attach to the lower portion of the pelvis to assist in our walking stride. The relative angle of the hamstring's attachment is a presumed measure of walking efficiency. Lucy's hamstring angle was longer than, but similar to, that of modern humans. The team noted the outward jut of the pelvic blade and the flatness of the bones at the base of the vertebral column (Johanson and colleagues had attributed this flatness to damage that had occurred to the fossil after death, not a feature of the living creature).

Stern and Susman noted that the end of the femur near the knee had a somewhat apelike appearance. And they found the knee itself to be quite unlike that of a human walker. They claimed that the *afarensis* knee was meant for substantial tree climbing. Finally, the feet of *afarensis* were similar to those of an ape. The toe bones were relatively long and curved, something that Johanson had interpreted as useful for walking upright over rocky terrain in which some gripping power might have been useful. Stern and Susman dismissed this suggestion.

Stern and Susman were not the first to make these interpretations. The French anatomists Brigitte Senut and Christine

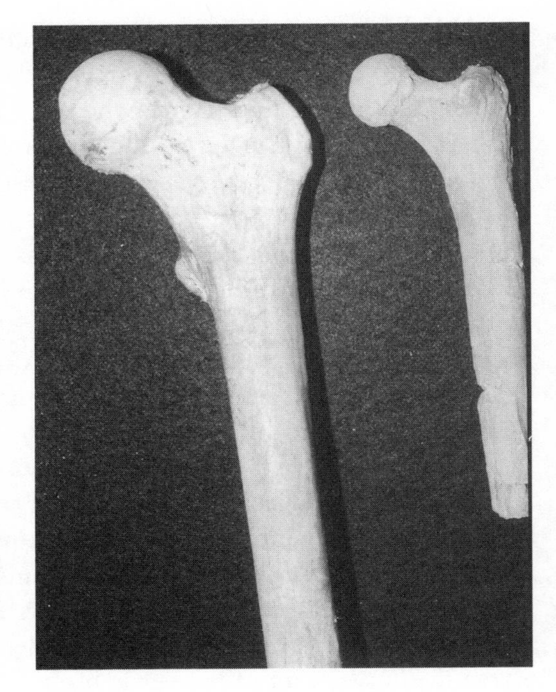

Lucy's femur (at right) is just a smaller version of a modern human femur, one clue that *Australopithecus afarensis* walked much as we do.

Tardieu had already concluded, based on separate studies, that Lucy retained the ability to climb well. And the University of Chicago anthropologist Russell Tuttle, who had helped to mentor Susman and Stern years earlier, had also looked at Lucy's skeleton and found evidence of a semi-tree-living lifestyle for *afarensis*.

At about the same time Jungers published a paper in the prestigious British journal *Nature*. He argued that Lucy's anatomy was primitive because her upper leg bones were short relative to those of modern people. Jungers stated that Lucy was not as agile a climber as a chimpanzee but that she also did not walk upright in the modern sense. Meanwhile, Bruce Latimer of the Cleveland Museum of Natural History and Owen Lovejoy claimed that the Stony Brook team had overlooked the degree of distortion in

Lucy's pelvis caused by the fossilization process. Their analyses supported the conclusion that Lucy was a "good" biped.

At this point in the debate, Johanson, Lovejoy, and their colleagues had the advantage. Johanson and White had disseminated the earliest scientific opinions about the specimen. Johanson published a popular book, *Lucy*, in which he detailed the discovery of the fossil and presented his own interpretation of her biology. White moved on to other research interests, and by the early 1980s arguing about Lucy was left mainly to Johanson, Lovejoy, and Latimer. Lovejoy, one of the foremost authorities on the biomechanics of walking, enjoyed an international reputation. Investigators often called him to crime scenes to help reconstruct the height and weight of a suspect who had left telltale footprints behind.

Johanson, Lovejoy, and Latimer counterattacked in the 1980s. Lovejoy accused the Stony Brook team of overlooking anatomical features that distinguished Lucy from other primitive humans: an expanded number of lower vertebral bones (six rather than the three or four that great apes possess) and a pelvic blade completely different in its shape than anything an ape possesses. Lovejoy's analyses showed that the dimensions of the supporting ridges and knobs surrounding Lucy's knee joint closely resemble those of modern people.

Bruce Latimer went beyond even Lovejoy's assertions: He argued that Jungers, Stern, and Susman simply didn't understand how natural selection molds a creature's anatomy and that therefore they misunderstood the meaning of Lucy's arm-leg proportions. According to Latimer, Lucy had already evolved the upper body of a human, and natural selection must have been acting to reduce arm length in Lucy's immediate ancestors, because *afarensis* had dropped tree climbing from its lifestyle. While lower limb changes in *afarensis* can be explained by natural selection's driving the rise of bipedal walking, changes in the upper body have only a minor effect on upright walking and so would not have been reduced as much as they were unless *afarensis* had given up tree climbing. Latimer acknowledged that Lucy may have climbed occasionally, but this seemed irrelevant to him, because she was not anatomically adapted to climbing.

The Stony Brook team responded predictably. Lovejoy, they said, misunderstood the anatomy of walking. What's more, they claimed that Johanson, White, Latimer, and the others had pushed their claims for full bipedalism by selectively picking and dismissing data points that did and did not fit their conclusions. Recall Lovejoy's reconstruction of Lucy's gait: He showed that the key evolutionary switch in roles occurred in her smaller gluteals — the gluteus minimus and medius — which converted from hip extenders (propulsion) to leg abductors (support and stability while standing). Susman and Stern said that Lovejoy's interpretation of Lucy's gait was based on a faulty analysis of the bones and muscles of walking.

The Stony Brookites also took a different research tack. They hooked a chimpanzee up to electromyography equipment, which measures the pattern of muscle firing during activities like walking. They found that the gluteals upon which Lovejoy had based his argument about Lucy's walking muscles seemed not to function in the way that Lovejoy had thought. Stern, who was the electromyography expert of the team, found that the lesser gluteal muscles were not aiding the chimpanzees in hip extension at all during bipedal walking. This added considerable fuel to the fire burning between the two intellectual camps.

This sort of nasty academic debate is often settled by the discovery of new fossil evidence. In the Lucy debate an old piece of evidence played a role. The most famous footprints in human history were made not in 1969 by Neil Armstrong in the lunar dust but nearly four million years ago in the soil of East Africa. Discovered in 1976 during a dig in northern Tanzania led by Mary Leakey, the 3.6-million-year-old tableau is preserved at a site called Laetoli. The ash of an exploding volcano had blanketed the landscape. Then it rained, dampening the ash and leaving droplet imprints. Ancient giraffes, elephants, and guinea fowl made their way across the scene, each leaving its footprints. Hominids strode across the ash too, leaving indelible tracks that were later covered by more falling ash, which turned into fossils through the eons.

A few facts about the fossils are clear. They are hominid. They were walking bipedally. Their feet possessed an arch like ours. Two, perhaps three, hominids were walking alongside each other. The latest forensic reconstruction shows that the three were four to five feet tall. One seemed to deliberately walk in the leader's footsteps. If a third individual was with them, its steps paralleled those of the others', and it changed its stride to keep pace. Whether they were a family with a child, or two males and a female, or any other combination will never be known. But the tracks themselves are of priceless anthropological value.

Anthropologists consider the Laetoli footprints, because of their age and location, to have been made by *Australopithecus afarensis*. Lovejoy, the footprint and gait expert, sees them as strong evidence that bolsters his claim that our ancestors walked much as they do today. Russell Tuttle of the University of Chicago has studied the tracks extensively and considers them so different from the bones from Hadar that they must have been made by another species of *Australopithecus* or even by an unknown species of genus *Homo*. He sees the stride of the Laetoli track makers as so modern that they look like the footprints that a little *Homo sapiens* would make in wet sand at the beach.

The Stony Brook group considered the footprints good evidence that *A. afarensis* was a "transitional biped" that was taking its first tentative steps out of the trees. Stern and Susman argued that having an arched foot does not necessarily mean that a creature is a hominid biped. Apes or people walking in sand can make a footprint with an apparent arch. The overall form of the feet, they decided, is not that of a fully bipedal walker. Stern and Susman granted that the Laetoli footprint makers had a big toe that was in the same line as the other four toes, certainly not divergent in the way that a chimpanzee's is. Stern and Susman also believed that some of the toe prints looked like what a chimpanzee would do in wet sand (it was partly to test these assertions that Susman undertook the infamous "clown shoes" study).

Tim White, who worked at Laetoli with Mary Leakey and published analyses of hominid teeth found at the site, has a different opinion of the tracks. He and his former graduate student Gen Suwa argued that the tracks were indeed left by *Australopithecus afarensis.* In a 1987 paper they dissected the Jungers-Stern-Susman argument, pointing out how subjectively one can find or dismiss features in a poorly made impression in ash or sand. I have a cast of one section of the Laetoli footprints in my own collection at the University of Southern California. It is, admittedly, only a roughly made replica and therefore lacks details of the original. But if I were walking down a beach and saw this trail of footprints in the wet sand, I would start looking around for a couple of children who had walked my way since the previous high tide. The footprints are obviously human; they don't remotely resemble an ape's footprint.

White and Johanson had long argued that the bones from Hadar and those from Laetoli, discovered in the days before the tracks made the place hallowed ground, are very similar. Meanwhile, there is not a shred of evidence at Laetoli that any other kind of more advanced human, such as Tuttle interprets, was the maker of the footprints. So we have three possibilities: The prints were made by a nearly modern person whose remains are not at Laetoli (Tuttle's idea); the prints were made by *afarensis*, which didn't walk like you or me (Stern, Susman, and Jungers); or they were the imprints of an *afarensis* best seen as a relatively modern biped, albeit one that may not have walked exactly as we do (White, Johanson, and Suwa). How do we resolve the debate?

. . .

The likelihood that the earliest hominids were fully upright walkers or bipeds who retained an ape's tree-climbing adaptations depends entirely on whom you talk to. The camps break into Johanson, White, Kimbel, Lovejoy, Latimer, the Israeli anatomist Yoel Rak, and a few others on the one hand, and the Stony Brook crew — Stern, Susman, Jungers, Fleagle, and their associates on the other. As in any political arena, linking up with one side or the other — being trained at Stony Brook versus Berkeley — carries some serious baggage. There's a domino effect too. Students trained at Stony Brook carry a Stony Brook worldview with them when they complete their degrees, settle at another university, and begin to train their own students. The influence is not as pervasive as it was at Berkeley back in the days of Sherwood Washburn, whose students comprised an enormous proportion of the influential human evolutionary scholars in the United States for a generation. Today we have too many research centers for that sort of intellectual monopoly. But the effect of a few scientifically and politically powerful mentors is still profound and influences generations of research, grant getting, and publishing, thereby setting the course for how we learn what we learn about human origins.

You may wonder how two groups of scientists, both composed of diligent intelligent people, could look at the same set of data and see it so differently. This is part and parcel of the doing of science. As the paleoanthropologist Katherine Coffing has argued, the reason that opposing research teams can't reach a consensus boils down to one of three possibilities:

One research team analyzed the data properly, whereas the other flubbed it.

Different research teams somehow interpret the same data differently.

Different researchers use fundamentally different notions of how evolution by natural selection works.

No doubt the first of these possibilities, that one team is simply wrong, happens as often in science as it does in life. This is what Stern and colleagues claim about Lovejoy. In a candid critique published in *Evolutionary Anthropology* in 2000, Stern examines the areas of his own analysis in which he might simply have been mistaken about research approaches, statistical tests, and basic premises. He finds a number of points on which his team might have erred and a number of others on which his colleagues may have erred. When many researchers are working on the same data and trying to answer the same questions, your competitors tend to find mistakes. You have a choice of how to respond to a colleague's criticism: You can acknowledge the possibility of an honest mistake and redo the analysis, or you can stonewall. The annals of academe are full of researchers who put pride before professionalism and paid the price. If you refuse to acknowledge an error and others have to point it out to you, in the end your mistakes will come back to bite you. Wrong results and faulty analyses get overturned all the time, to the great embarrassment of researchers.

The second possibility, that different research teams can see the same data in two different lights, is also common. Stern points out, for example, that his team disagreed vociferously with other researchers about the finer points of Lucy's locomotor anatomy, and in some cases the language that both sides used belied the fact that the analyses had produced essentially the same results. For instance, Stern and Susman's analysis of Lucy's heel bone essentially matched the results obtained by a French bipedalism researcher named Yvette Deloison, but you would never have known that by reading their respective papers. Latimer and Lovejoy looked at the same toe bones in which Susman and Stern saw a pronounced curvature and saw virtually no curve. Blinding yourself to the possibility that an archrival, or even just a colleague, could be right (or, worse, may have scooped you) is a common self-delusional strategy in science.

Coffing's third point was the possibility that scientists may reach different conclusions about the early hominid fossils, because the scientists are operating with different fundamental understandings and interpretations of how evolution works. She finds this case the most compelling. Scientists are trained in different schools of evolutionary thought. This, added to the preconceptions that we all bring to our research, is the recipe for a creative soup of ideas about how evolution works, but it may also prevent us from sharing the same paradigm. The paradigm, or working conceptual model, of natural selection is constantly being tweaked this way and that by new research and new theories. In Lucy's case this might mean that the Stony Brook scientists and the Berkeley-Cleveland group bring different fundamental assumptions to the table, and these deeply held views influence the ways that they interpret the same bits of information.

Can we find clues to these "deep assumptions" in the writings of the two groups? Latimer's publications thoroughly examine how natural selection works and how it is revealed in the fossil record. He stresses the role of directional selection — natural selection that pushes an organism strongly one way or another. (Stabilizing selection, the converse of directional, is the evolutionary force that keeps animals looking essentially the same as their parents without some intervening reason to change.) Lucy's behavior was different from that of her predecessors, and this drove changes in her anatomy. Implicit in Latimer's thinking is that human fossils never quite represent a state of equilibrium in evolution.

The crucial matter is how one views optimality. Natural selection molds a species by conferring greater reproductive success on those members who do something a bit better than their neighbors. An ape that walked "better" (that is, faster, farther, or more safely) on less energy by standing upright occasionally would have a bit more energy to invest in finding a desirable mate or nutritious food. If he or she left offspring, the genetic complex that coded for the change in posture — and that capitalized on the change in behavior that preceded it — would spread. Over many generations the species would change genetically, anatomically, and behaviorally, perhaps so much so that millions of years later we would see this change in the fossil record and assign different names to the different forms. At each stage of the process the evolving ape-human was presumably better at what it did than its neighbors, or else the trait would not have been perpetuated. In other words, we assume that natural selection molds optimal creatures and that all traits of an organism are optimal at all times.

But how can this be? Doesn't the nature of changing from one state to another imply that, along the way, not every intermediate state was perfectly designed? If you begin with a gas-powered car and try to re-engineer it into one that runs on electric fuel cells, won't the experimental models be too much of one but not enough of the other? Natural selection doesn't seem to work this way. At each stage, if the design is wrong, the evolutionary line dead-ends. This is part of why evolution by mutation and natural selection is so slow and inefficient.

But Latimer does not assume optimality in the course of human evolution. He looks at Lucy's skeleton and sees some features — her upper arms, for instance — that are not very competently designed for climbing. Therefore, he assumes, she probably didn't climb. But how do we know that Lucy didn't climb even though she was poorly adapted for doing so? Because behavioral change precedes anatomical change, *afarensis* must have stopped climbing and begun walking upright on the ground long *before* its anatomy fully reflected this. But this would have happened during a narrow window in time, after which anatomy began to change and "catch up" with the behavioral change, so it may not be very visible in the fossil record.

The sides of the debate are starkly drawn. Lucy was a full biped, say Lovejoy, Latimer, and the others. Lucy was a transitional biped, still a climber, say Stern and Susman. The debate resembles the plot of the Japanese film *Rashomon*, in which a sinister crime unfolds in full view of several eyewitnesses; the police investigation falls apart when the witnesses all reconstruct their limited perspective of the events differently. In the case of *afarensis*, the same evolutionary facts are regarded differently by witnesses viewing the bones from irreconcilably different intellectual angles.

But the Stony Brook team erred badly when it adopted the implicit view that Lucy was a "transitional" biped. *Transitional* is a term that makes sense only in twenty-twenty hindsight, and that's not the way to understand the course of evolutionary change. In fact, it's an excellent way to misunderstand it. Meanwhile, Johanson and company also painted themselves into an intellectual corner, that of full, habitual bipedalism. Once wedded to that view, acknowledging even occasional tree climbing means losing the debate and that the Lucy-as-climber advocates will drive the academic knife home. This is a main reason why the Johanson-Lovejoy camp has claimed since the 1970s that Lucy was a ground-living biped.

ONE OF A KIND

A scientifically sound way to reconcile the evidence may exist, however. Most of the research on Lucy's skeleton was conducted in the late 1970s through the 1980s. The view of Lucy and what she tells us about bipedalism has changed a great deal since then. A few researchers have stepped up to show that Lucy and her kin possessed adaptations that needn't be seen as either like modern humans' or like chimpanzees'.

Lucy and her cousins were unique, without good modern analogs. Their way of getting around, while obviously bipedal and also showing some evidence of climbing, was not transitional at all but rather a third state. The anatomist Yoel Rak, a close colleague of Johanson and White's, has analyzed Lucy's pelvis and concluded that her gait was not an intermediate stage but a solution to the compromise anatomy that a new biped must have had. The widening of her pelvis carried with it anatomical disadvantages and advantages. Her relatively short legs meant that as the center of gravity moved upward as her ancestors became bipedal, natural selection compensated by widening the *afarensis* pelvis and lengthening the neck of its femur. The effect of these compensations was to allow bipedal walking without seriously compromising the energetic cost of travel.

The key to Rak's ideas about Lucy is his argument that her pelvis is not an intermediate stage but a solution to the compromise anatomy that a new biped had. Meanwhile, even those who see climbing adaptations in Lucy's bones don't agree about how she could have climbed. The anthropologist Laura MacLatchy of Boston University, for instance, argues that *afarensis* may have climbed very differently than any modern primate.

Patricia Kramer and Gerald Eck of the University of Washington have produced further evidence that Lucy was unique. In addition to her training in human evolution, Kramer is a civil engineer with the Boeing Corporation, accustomed to applying engineering know-how to transportation problems. Most researchers have been hidebound enough to insist on using the environment in which our own form of modern walking arose when comparing all walking in our ancestors. When we do this, Lucy will always look like the poor cousin, which sends us scampering to call her "transitional," "on the way to being a biped," and other such silly epithets that are used repeatedly in human evolutionary research. Kramer and Eck showed that Lucy was neither a modern biped nor an inefficient, partial one. Instead, the best way to view *afarensis* is as a biped adapted to a very different sort of bipedalism, and for wholly unique reasons, than our own. According to the researchers, Lucy was highly adapted to slow short-distance walking. Her anatomy was perfected for a set of ecological pressures that mandated a particular style of locomotion. This likely would have had to do with the nature and distribution of foods that she ate.

HABITATS FOR HUMANITY

Fossil remains of Lucy and other early humans offer important clues to the environment in which they lived, and also their — and, by implication, our — societies. Our hidebound thinking that we are the endpoint of evolution has led us to our deeply held view that the key phases of human evolution occurred on the savanna. The notion that humans emerged in flat open country has a long history. It predates, in fact, the discovery of any fossils or environmental data that could provide evidence either way. When Dart discovered the Taung child in a nearly treeless veldt in South Africa, it certainly made sense to suppose that the creature had in life been utterly divorced from the forest, except as a place to forage occasionally. As with many other aspects of the emergence of bipedalism, however, our early simplistic view is giving way to a more complete and realistic picture of early hominid life.

The idea that we evolved in grassland has been profoundly influential in the history of ideas about who we are. Many studies have shown that people feel more psychologically comfortable in an open parklike environment with scattered trees than in any other setting. One study showed, for instance, that children prefer savanna-like landscape photos, even though none had ever been to a savanna. This supposed predisposition was thought to account for the way our parks are landscaped. The evolutionary interpretation of this preference for a parklike landscape may have some psychological foundations in the landscape in which our brains spent most of their evolutionary history. Grasslands with scattered trees offer good visibility for both game animals and approaching dangers, plentiful food, and refuge from harm. A generation of researchers relied on the assumption that extinct humans had made their living by scavenging or hunting game on the savanna, the so-called Serengeti model.

It all makes perfect evolutionary sense. It's also completely wrong. Although some early human landscapes were seasonally dry savannas, others were emphatically not. Some of the earliest evidence of humanity comes from habitats once thought to be arid but now believed to have been forested or swamplands. No doubt each species had its own preferred place. Kaye Reed, an Arizona State University paleoanthropologist who reconstructs the environments of early humans, examined the ancient communities of animals that shared the world with emerging humans. Numerous other studies had worked from the assumption of evolutionary continuity: If an ancient habitat held the remains of fossil antelopes whose modern counterparts graze grass, the extinct species also were grazers and probably lived on a savanna. Reed took this research to a new level, examining the anatomical adaptations of each fossil group rather than assuming evolutionary continuity. She found that the habitat of most of the earliest australopithecines was generally well-wooded regions with lakes and rivers. Later, the robusts preferred habitats that included wetlands. Meanwhile, the archaeologist Martha Tappen of the University of Minnesota studied modern scavenging carnivores — hyenas and lions — in East Africa and found regions of forest-grassland mosaic to be just as probable a setting as drier and more open savannas for our ancestors to make a living. Even the Laetoli footprints site, long thought to be the most arid of early human environments, is now thought to have been less than a full-fledged savanna, according to Peter Andrews of the British Museum. Only beginning with the genus *Homo*, about 2.5 million years ago, did humans spend more time on the open plains of Africa.

Thinking of a single habitat as the cradle of humanity is silly. Even at the earliest stages hominids succeeded because they were ecologically adaptable generalists. Our plasticity is what allowed us to survive and prosper as a lineage.

LUCY'S LIFE

> Scenario A: Lucy was a terrestrial, highly refined biped. She lived in a social group with a number of males and females, and she mated promiscuously, as it suited her.
>
> Scenario B: Lucy was semi-arboreal, climbing trees with agility. She lived in a monogamous pair bond with a male of her kind.

We have no evidence that either of these views of Lucy's life is correct. Both are speculative, and both are possible. Being terrestrial or arboreal is not necessarily connected to a particular mating system, either. But how Lucy lived — her behavior — is central to the debate about her anatomy.

And the argument about Lucy's anatomy is, to say the least, contentious. But her behavior molded her anatomy. Reconstructing behavior will always be speculative, because social behavior doesn't fossilize the way that tibias and tailbones do. We can reasonably infer some important features of australopithecine society, courtesy once again of the best tool in the toolbox, Darwinian theory. Darwin put together the theory of sexual selection a decade after he showed the world how natural selection works. The competition among males for the opportunity to mate with females is one lane of an evolutionary two-way street. The other lane is a female's choice of her preferred male. Females are a driving force in the evolutionary process, because they decide which males' genes enter the next generation in abundance. Females who choose to mate with large-bodied males will ultimately be the arbiters of a future population of large males and smaller females. Because males compete, sometimes fiercely, for the chance to mate, sexual selection drives changes in body size, musculature, and all manner of sex differences in anatomy and behavior.

Australopithecus afarensis males were a good deal larger than females in body size and weight, a size difference that was greater than that between modern men and women. The male *afarensis* had a far larger and stronger skull and canine teeth, which gave him a macho profile. But Lucy was the smallest *afarensis* specimen discovered; comparing her to the largest *afarensis* specimens led some researchers to argue erroneously that early hominids were extremely *different* in male-female body size. The sex difference was dramatic enough that early researchers initially debated whether two species of australopithecines existed among the Hadar fossils. White and Johanson argued hard and won that debate, showing that the best statistical match for the size range of the Hadar sample was comparison with modern ape species.

The pronounced sex difference in anatomy among Lucy and her kin means (assuming that extinct animals followed the same evolutionary patterns that modern animals do) that *Australopithecus afarensis* lived in groups consisting of at least one male, perhaps several, and a larger number of adult females and their offspring. Although chimpanzees or bonobos are not perfect reference points for extinct hominids, they do offer some lessons about the range of possibilities for the lifestyle of *afarensis*. Both chimpanzees and bonobos live in highly fluid societies that feature much strategic promiscuity. Males are territorial about community boundaries, chimpanzees lethally so. Foraging needs and female reproductive cycles dictate grouping patterns: Female chimpanzees prefer to forage alone or in small parties. Female bonobos are more sociable but still spend more time in smaller groupings than males do. Females of both species migrate to other communities at or after puberty and settle in as breeding residents someplace other than where they were born and raised. Males of both apes cooperate in endeavors that range from monitoring territories to trying to control females. Male chimpanzees hunt other mammals avidly in male coalitions. Among bonobos, females also form alliances among themselves for the purpose of evading male domination.

We don't know whether males of her species fought over Lucy, but it seems likely, because male competition for females is a prominent feature of the living great apes. Gorilla, chimpanzee, and bonobo females are all quite strategic when it comes to mating; they are active tacticians who seek the best males and perhaps the best genes to bestow upon their offspring, wherever they can find them. Female apes are not passive recipients of male mating ambitions, even though it took decades for primatologists to come around to that realization. Orangutans, with an even greater degree of male-female size disparity, live somewhat solitary lives, though many scientists suspect the recent ancestors of the orangutan did not.

What can we extrapolate from these apes to help us get a more accurate picture of our ancestors' lives? Richard Wrangham, of Harvard University, and others have argued that the earliest hominids probably lived in closed groups and that, as we see in chimpanzees, male kin groups were the glue. The more recently studied bonobo tells us that females may have played a more important role in australopithecine groups than we would think based on chimpanzee behavior. We'll never know which ape is the better model for earliest humanity, although I lean strongly toward the chimpanzee. Like the earliest humans, the chimpanzee occurs in a wide variety of habitats over a huge geographic area, from grasslands with patches of woodland in far eastern and western Africa to the dense tracts of primeval lowland rain forest in the

Congo basin of the center of the continent. This ecological diversity and wide geographic distribution dwarf the bonobo's single habitat of lowland rain forest in an area below the loop of the Congo River in the Democratic Republic of Congo. The chimpanzee's larger range is matched by a dramatically greater tool technology, with cultural traditions of tool use that range from stone hammers to twigs used to fish for termites. Bonobos, for all their fascinating sociosexual diversity, are minimalists when it comes to making and using tools in their natural habitat.

The chimpanzee's distribution and habitats may resemble those of the early australopithecines. *A. afarensis* and the other species have been found mainly from the Rift Valley of East Africa, that flat expanse of savanna that lies in the rain shadow east of the mountains of the Great Rift, to the grassy veldts of southern Africa. However, these areas are probably not the only homes of the earliest hominids but only those that contained conditions favorable to fossil preservation. The recent discoveries by Michel Brunet of what may be hominid remains from arid Chad, far to the west of any previous australopithecine finds, tell us that early hominids may have lived across Africa, perhaps in forms quite different from those that we know of.

So a reasonable retrospective guess for the mating system and grouping pattern of, say, *Ardipithecus ramidus* or *Australopithecus anamensis* would be male-bonded groups that defended large territories in forest-grassland mosaics. Within these territories females and their children lived together, either in small parties or perhaps larger groups, emigrating from them at puberty. When males weren't trying to ward off intruders, they competed among themselves to mate with the females. They ate a diet of fruits and leaves, supplemented by insects and whatever red meat they could scrounge or kill in the form of small game animals.

If this sounds strikingly like a chimpanzee to you, that is no coincidence. Modern chimpanzees are by far the best analog we have for figuring out what Lucy was all about. But some scientists are not happy with this approach. Some theorists have criticized the use of "models" such as chimpanzees, saying they mislead us from the whole range of evidence about australopithecine behavior. These anthropologists advocate making use of information from a wide variety of fields. Yet these theorizers invariably conclude that the last common ancestor of humans and chimpanzees was indeed very much like a chimpanzee.

As my colleague John Allen of the University of Iowa and I pointed out in a 1991 paper, even the best attempt at a conceptual model for early humans is, in practice, just a layered set of referent models — take the best bits of a chimpanzee, a bonobo, a gorilla, and some other primates, roll them into one, and voilá — you have the sort of portrait of an early human that seems plausible to scientists of all stripes. This highlights how hard it is for us to escape the mental constraints of the present.

NEST EGGS

A far safer parlor game than the guesswork about australopithecine social behavior is their ecology. After all, while social behavior doesn't fossilize (although male-female size differences may fuel speculations about mating systems), the animal's relationship to its physical environment may be revealed more clearly. It's hard to imagine how an early apelike hominid could have survived the African night unless it slept in trees as modern chimpanzees and bonobos do. Another pretty reasonable assumption is that those earliest humans used tools, because modern apes have impressive tool technologies. Some researchers have taken these premises to heart and tried to reconstruct australopithecine behavior based on the distribution pattern of chimpanzee nests and tools in modern African landscapes.

Jeanne Sept, an archaeologist at Indiana University, studied chimpanzee nests with an archaeologist's eye and found that they occur in patterns similar to the concentrations of artifacts, such as stone tool accumulation, in the fossil record. Perhaps, she reasoned, ancient people bedded down in ways that are analogous to how great apes sleep today. Great apes generally make a new nest every night, climbing into the treetops to break branches inward toward a central bowl that becomes a very comfy-looking bed (gorillas are an exception, typically nesting on or near the ground). But chimpanzees sometimes return again and again to the same nest tree or grove of trees, perhaps because it is conveniently located for the next morning's breakfast of fruits or maybe because it's just a comfortable tree for bedding down. In other words, even though we usually define a home base as a uniquely human feature, some apes also tend to return to favored places to sleep. Sept reasons that as humanity evolved, the use of home bases became a more and more critical piece of our ancestors' lifestyle. Home base became a place where the activities of daily life — sleeping, eating (food may have been carried back there), tool manufacture and use — were concentrated. Centers of activity leave their mark on the landscape, and that is what archaeologists study when they try to reassemble the distant past.

What trace, you may ask, could possibly remain today of leafy nests and stick tools made five million years ago? In West Africa chimpanzees use stones as hammers to open nuts. Although they are not intentionally carved or flaked to make them more effective tools, through repeated banging the edges take on distinctive wear patterns that a sharp-eyed archaeologist can see eons later. And the gathering of the tools in itself changes the natural distribution of stones on the landscape and may be discernible in the present. Such chimpanzee-created clusters of tools are archaeological sites in the making, and some scientists, notably the French researcher Frederic Joulian, have begun to treat them as such.

Even nests may leave some trace. Barbara Fruth and Gottfried Hohmann of the Max Planck Institute for Evolutionary Anthropology found, while studying bonobos in the forests of Congo, that trees that are used for nesting suffer distinctive deformities as a result of having their branches broken into a bonobo bed. Once Fruth and Hohmann had trained their eyes to look for these distinctive breakage and regrowth patterns, they were able to identify trees that had been used by sleepy bonobos years earlier. Thus, even forests where the apes may be extinct would give up some details of what once was, if the trees are still standing and you know what to look for.

Lucy will always be a cipher to science, a specimen to which many will look for answers. We will probably never agree on the details of what she tells us, and it is not even clear that *Australopithecus afarensis* was a direct ancestor of modern humans and not the ancestor of the robust hominid lineage or another, yet unknown, lineage. The big picture is clear, however. Bipeds arose early and quickly in human origins and almost certainly included not only those that eventually led to modern humanity but many others barely known or yet undiscovered. All evidence points to a form of bipedalism in which natural selection tinkered and toyed but never tried to create "the perfect biped."

6
WHAT DO YOU STAND FOR?

THE KALAHARI DESERT OF southwestern Africa is a harsh place to live. The climate is so inhospitably hot and dry that most animals must live underground during the day. When they do come out to forage, they must cope with a variety of predators, from poisonous snakes underfoot to birds of prey overhead. Survival is determined by intense vigilance and some luck. In this environment a small mammal is a savory snack for a wide range of larger creatures.

One such small mammal is the meerkat, a diminutive member of the mongoose clan. Looking like a cross between a tabby cat and a rat (a meerkat had a starring role in *The Lion King*), these small carnivores live out their daily lives while coping with the risk of death-by-predator every waking hour. Meerkats have evolved a unique strategy for dealing with this threat. When meerkats leave their burrows in the early morning hours to forage, one or more of their group stands sentinel on the raised rim of earth above their subterranean home. But the way they stand has attracted the attention of human evolutionary scholars. Meerkats stand upright while guarding their kith and kin. As though they are auditioning for the Coldstream Guards at Buckingham Palace, meerkats stand on two legs — alone, in pairs, or trios — their backs arched rigidly to maximize their ability to scan their surroundings. The appearance of an eagle elicits alarm calls from the sentinels and a hasty retreat to the burrow by the entire clan.

Meerkats stand only a foot or so tall, but scientists believe that those few extra inches of height that they gain by raising themselves up momentarily give them an edge. Bipedal posture for vigilance is also one speculation about why our ape ancestors decided to stand up. Unlike the meerkat example, in which sentinel behavior has been shown to serve as a survival benefit for kin of the sentinel, human evolutionary experts can only imagine what the benefits of bipedal vigilance would have been for our ancestors five million years ago. So standing up to be vigilant is a theory of human origins, but it is not testable unless we look outside the primate order and see its value in the little meerkat.

When we want to figure out why the earliest humans may have stood up and walked, we usually turn to other animals. But examples from the animal kingdom, including the meerkat, are fraught with problems. Why, for example, would an early hominid need to stand on two legs habitually, when it serves the meerkat well to stand only on occasion? The meerkat story provides a useful, albeit limited, analogy, because it fits a particular story about human origins — that we had a need to see what was happening in the distance.

Theories about human origins are supposed to be rigorously scientific, not fanciful stories. Yet getting anyone in the scientific community to accept your theory is difficult unless it takes the form of a story. This results from the narrative nature of our souls. I can stand in front of a lecture hall full of eager undergraduates, but only when I switch from explanation by fact giving to explanation by narrative — telling an anecdote about my life of studying wild primates — does every head in the room snap to attention. We're better listeners when the style, not just the content, hooks our interest, as Misia Landau articulated so well. Despite the aura of scholarship that many scientists hide behind, thinking up a new and exciting theory usually requires not only data and a good imagination but also some clever marketing ploys.

What separates a solidly conceived, easily accepted theory from one that will fall flat is straightforward. It must accord well with the cutting edge of the field. Every new fossil that's discovered leads to new theory making, because the facts that we hang theories upon have changed. Second, it has to be internally consistent; the pieces must not conflict with each other. If you achieve internal consistency and use well-established facts, chances are good that your theory will have a long lifespan, although its luster may tarnish when exposed to new facts discovered years later.

Another issue is the paradigm. Every scholar, consciously or not, approaches an intellectual problem with a certain paradigm — a way to view your research that is inherited from the generations that came before you. We have already seen the tectonic crunch that occurred when competing paradigms about Lucy's way of walking collided. The late philosopher Thomas Kuhn offered a conventional wisdom about paradigms: An old paradigm is not replaced by a new one because of accumulating facts that run counter to it. Instead, a paradigm stretches to encompass new evidence until it creaks and groans and the fit between fact and theory becomes utterly unlikely. Then someone comes along — like Einstein challenging Newton — and the paradigm bursts like a bubble. When the dust clears, a whole new paradigm is in place, a new way of understanding the world that everyone would have dismissed yesterday but today seems hard to believe we could have lived without. The new paradigm is often so brilliantly elegant and simple that its disciples can't believe that they were so stupid that they did not enunciate it themselves. Such was the earliest reception of Darwinian evolutionary views by nearly all his fellow scientists back in the 1860s.

So far, I've tried to convince you to throw away the idea that humans evolved from one bipedal ancestor — the "missing link" — that improved until it became *Homo sapiens*. I want you to adopt a new worldview that does not include the concept of progress in early bipeds. We have absolutely no reason to think that bipedalism arose for one reason only or to believe that the same factors that drove the origins of bipedalism later molded it into greater efficiency. Only magical thinking allows us to believe in old simplistic linear models.

For the sake of putting forth a working idea about something poorly understood, we make certain assumptions and then build on them. One time-honored way to figure out the behavior of a long-extinct creature is to assume that it resembled a modern analog. As I've noted, the chimpanzee is the most widely invoked example of what an early hominid may have been like. Back in the 1960s Sherwood Washburn and his student Irven DeVore relied heavily on the baboon to persuade anthropologists that they should stop sitting in their dusty labs measuring skeletons and get out into the African bush to watch living primates. Some scientists prefer the bonobo to the chimpanzee as a model. But using an analogous species may be a simplistic approach to understanding the origins of humanity. Fixating on any one modern species as a good model of what Lucy might have been like may, in relying on a few superficial similarities, cause us to ignore the rich diversity of adaptations that our ancestors probably had. But as a starting point, ignoring chimpanzees, bonobos, and other primates as models for ancient humans is nearly impossible. The primates may not be exactly like their ancestors and ours, but they are the next best thing. Most often, as we shall see, when scholars have assembled a broad range of facts to build a conceptual sculpture of the earliest hominid, the end product looks very much like a chimpanzee. This may be because we're all raised on a diet of great apes, from gorillas in the zoo to images of chimpanzees on television. Breaking out of the mind-set that early humans looked and acted like chimpanzees is nearly impossible. We should be grateful that we have the apes around, because without them, adding flesh and hair to Lucy's skeleton would have been complete guesswork.

Consider some of the most prominent theories about how and why our ancestors adopted bipedalism. Models of human origins have invoked key mechanical changes, physiological changes, and other behaviors, such as being able to see over tall grass and carry food. And many theories incorporate two or more of these categories. The problem with invoking a single behavior, cultural tradition, or technology is that the fossil record almost never provides any physical evidence of the behavior. For example, Nina Jablonski of the California Academy of Science and her colleague George Chaplin have published a theory of human bipedalism that says the need for early humans to quell group disturbances pushed them to stand up and make impressive displays. Only highly circumstantial evidence for the theory exists, however, because such behavior leaves no trace in the fossil record. Like many other theories of human evolution, it is mainly a well-constructed speculation. Raymond Dart long ago theorized that the biggest benefit of early hominids' standing upright would have been to see predators coming over tall savanna grass. The basic problem with this idea is that, as anyone knows after becoming stuck behind a crowd at a parade, you don't have to stay tall for more than a few seconds to get a peek at what's coming. Dart's is an appealing story without supporting evidence.

Another widely believed theory of human origins is a well-constructed speculation: the aquatic ape. Although nearly all the scientific community has dismissed it, the aquatic ape hypothesis enjoys wide public support. Elaine Morgan, building on an idea put forth decades ago by Alastair Hardy, claimed in a series of books that humans passed through an aquatic stage in their evolutionary history, one that has left detectable traces in our anatomy and physiology. According to Hardy and Morgan, our ancestors must have lived at the seacoast or on lakeshores at some critical juncture of evolution. Morgan cited our relative hairlessness, like that of dolphins and seals; our high level of subcutaneous body fat, which provides buoyancy; and our controlled breathing — perfect to prevent drowning during long dives to the bottom of a lake or other body of water. She points to our odd thermoregulatory capacity, our manual dexterity (perfect for shucking oysters), and even to our bipedalism itself (ideal for wading in the shallows and not bad for a swimmer, either).

The aquatic hypothesis is, like many more reputable theories, a pile of premises stacked on top of one another. The problem is, the premises are all wet. The biologist John Langdon of the University of Indianapolis has shown that virtually every one of Morgan's claims about the aquatic adaptations of humans is spurious. Aquatic animals have no greater tendency to be hairless than terrestrial mammals have to be furry — consider elephants, rhinos, and pigs. The idea that our hands evolved to eat shellfish is obviously a just-so story. No doubt you could sit down with a blank piece of paper and come up with a dozen equally clever speculations in the next few minutes.

The aquatic hypothesis is an example of the kind of house of cards that students of human evolution should avoid. But it does give an important lesson. It reminds us how important internal consistency and storytelling are to theory building. For better or worse, the most influential theories of human bipedalism are the "prime-mover" theories. These models for how bipedalism arose use one key feature as their fulcrum, then draw together information from a wide range of sources, and make a comprehensive, one-size-fits-all argument for where bipedalism came from. Of course, whenever you start heaping one premise on top of another, you're asking for trouble, and, as we will see, most prime-mover theories fall apart under their own weight. They tend, however, to be highly influential, even becoming known by the names of their creators; hence, the "Lovejoy model," or the "Jolly model" of bipedalism.

If a theory for bipedalism is to hold its own, it must have a solid basis in the data. The earliest theories about human evolution were, as we saw in chapter 1, rooted in the notion that using tools was fundamental to making us human. Darwin thought of it first, and for the rest of the nineteenth century and through the twentieth, our thoughts about the importance of tool use have ebbed and flowed. Sherwood Washburn tried to revive interest in tools as a prime mover of human evolution, but the evolutionary disconnect between tools, which appeared at 2.5 million years, and the shift to bipedalism, which began three or so million years earlier, is crystal clear.

Of the other prime movers, a few have held sway. Raymond Dart himself theorized that the biggest benefit to early hominids' standing upright would have been their ability to see predators coming over tall savanna grass. Among the problems with this idea is the recent realization that the earliest stages of human evolution likely took place in a forested environment rather than in open country.

When a chimpanzee or bonobo stands upright, it sometimes does so in order to carry something. They don't do it for long distances because chimpanzees, as we have seen, are notoriously inefficient at long-distance walking on two legs. If they must carry something for more than a short distance, chimpanzees commonly tuck an object, perhaps a rock or a piece of meat, into the notch of their groin and carry it on all fours that way. So theories about the benefits of standing up to use your hands for carrying suffer from a lack of comparable behavior in the living apes. Carrying is nonetheless an attractive prime mover for many theory makers. You can carry food from where you found it or caught it to a safer place for dining, and you can carry tools to the spot where you need them.

Perhaps the most talked-about theory since the 1980s is one described by Owen Lovejoy that incorporates food carrying, monogamy, and the origins of the human female reproductive system. Writing for the prestigious journal *Science* in 1981, Lovejoy began with a well-established premise. The diversity of types of fossil apes, he observed, had been declining since well before the emergence of the hominids. At one time, as we've seen, the tropical forests of the earth were dominated by a rich diversity of apes, comparable to what we see today among monkeys. But in the modern world, even without forest cutting and poaching by people, many kinds of apes may have been headed for extinction. The four living species are but a pale remnant of their former species-rich glory. Why?

Lovejoy claimed that this reduction occurred because of the pathetically slow reproductive rate of the great apes. None of these animals gives birth as often as the two-year interval in many human societies; the interval between successive gorilla births is about four years, four to five years for chimpanzees and bonobos, and as long as six to seven years for orangutans. If a female orangutan has her first baby when she is sixteen and her last at thirty-six (typical in the wild), she has time to bear and rear only about three babies (a mother baboon or rhesus monkey has infants often enough during a shorter reproductive lifespan to more than double that rate). Monkeys simply outbreed apes, and, according to Lovejoy, this consigned the fossil apes and their descendants to the scrap heap of evolution.

More important, Lovejoy decided that the entire human lineage, as an offshoot of an ancient ape lineage, would have gone the same dismal reproductive route, unless natural selection molded adaptations that allowed our direct ancestors to compete successfully with monkeys. This breed-or-die dilemma came about five million years ago, when a drying climate was breaking the great tropical forests in East Africa up into a quilt of forest and grassland. Fruit trees that fed emerging hominids became fewer and farther between, and any ape or human who wanted to find good food had to cross ever wider stretches of open country.

According to Lovejoy, into this circumstance walked the earliest hominids. Remember that Lovejoy has long been one of the fiercest advocates of highly refined bipedalism in even the earliest bipeds. He has seen the emergence of bipedal gait as an adaptation to crossing wide patches of savanna in search of fruit and perhaps also to increase the amount of meat in the diet. He has gone on to point out that most supposedly uniquely human traits are not unique after all. Other primates have opposable thumbs and many of the other anatomical traits that we once thought to be solely human features. Tool use began much later than the rise of bipedalism. The one trait that most crucially set our first ancestors off from their simian cousins was, according to Lovejoy, female reproductive physiology.

You've seen the pink balloons that hang obscenely from the rear ends of female chimpanzees. These sexual swellings are vivid billboards of sexual availability, advertising that the bearer is either ovulating or otherwise eager for a male's attention. They occur in lurid form in chimpanzees and bonobos but are notably absent in women. Lovejoy saw the possibility that a link existed between our no longer advertising ovulation and other traits that were emerging as humanity evolved. Female protohominids, he decided, had to cope with a patchy environment, widely dispersed food, and males eager to mate with as many females as they could. This was not a winning game plan for a female, because it meant that she would be abandoned for all but a few fertile days of each monthly cycle. But if the female protohominid began to conceal her reproductive state by losing that headline-grabbing swelling, a male would have a far greater incentive to hang around, perhaps even to offer his services to her in the form of providing food.

This is where Lovejoy saw the tie-in with bipedalism. As forests contracted, males had to walk farther to find food to carry back to the females whom they were covetously guarding from the attentions of other males. Two-legged walking raised the energy efficiency of walking and enabled the male to carry food in his arms. Back at home, the female's physiology, ratcheted up a notch with the extra nutrition that she received from her now-attentive mate, could produce more offspring. The interval between births shortened, and the emerging hominids not only staved off extinction at the hands of monkeys but invaded a new grassland niche and ultimately conquered the world.

Lovejoy's theory was certainly ambitious. It tries to tie together information on ancient climate, anatomy, and reproductive physiology and speculation about behaviors such as carrying, fruit eating, and human mating. But many experts considered the theory as riddled with holes as with new ideas. We now think that critical stages of human evolution took place in the forest, not on the savanna. And contrary to Lovejoy's assumption that the ancestor of chimpanzees and humans had a huge sexual swelling that our own ancestors lost, we have every reason to think that chimpanzees and bonobos have evolved the huge sexual swellings, not that we lost ours. Perhaps most perplexing, Lovejoy assumes, in the face of overwhelming evidence to the contrary, that the ancestral, "natural" mating system of humanity was monogamy. Much evidence points to ancestral hominids as having a polygamous mating system much like that of modern chimpanzees.

Despite these problems, Lovejoy's model attracted a great deal of attention because the story is compelling. Other theories, perhaps more fundamentally sound but lacking in the narrative qualities with which Lovejoy imbued his, have fallen by the wayside since the 1980s, even as the debate about Lovejoy's has continued. The story features an underdog (the emerging ape-human) that must overcome villainous forces (a slow reproductive rate and a changing climate) to achieve its goal, world supremacy. And in the best of Hollywood endings, it does.

THE FIRST STEPS

There is, I believe, a far more plausible theory for the origins of bipedal walking. To understand it, we have to shake loose the notion that it had to happen for *one* reason or in *one* step. Like everything else in our anatomy, the components came together at different times and for different reasons. Rather than see the first steps as a clean break with the past — entering a new habitat or beginning a new way of life — it makes much more sense to view bipedalism as emerging when an ape that was *already* doing something began to do it more often; later the behavior became set in the stone of their anatomy. Theories that advocate sudden, radical departures from an ape way of life toward one that is very human receive little support in the scientific community. This spells doom for the carrying theories, because apes simply don't carry things, nor do they walk upright in order to do so. On the other hand, once the ape had evolved into a bipedal hominid, it may well have used its newly freed arms to carry tools, babies, and the like. Thus it follows that any theory of human origins is suspect if it has at its heart a very human activity.

If our first bipedal steps were taken in the shade and safety of the forest and not in the carnivore-infested glare of the African savanna, how can we account for the need to be bipedal at all? Like walking itself, the answer is probably far less sexy than the theories that get all the publicity. The key to understanding the origins of bipedalism is to forget the notion that it had to happen for *one* reason or in *one* step. Bipedalism is much more like a complicated piece of artwork — a mosaic tile — that was assembled over several million years.

An essential reason to stand and walk upright must be intimately tied to day-to-day survival and reproduction, that is, eating or mating. So we can eliminate imaginative scenarios such as carrying or seeing over tall grass. How being bipedal might enhance mating success is difficult to imagine, except perhaps by allowing high-ranking males to engage in two-legged dominance displays. But this wouldn't explain why bipedalism arose in females too. Given the dramatic changes that the body had to undergo, and the survival trade-offs involved, an evolutionary pressure that applies equally to both females and males is much more logical.

Finding and eating food are what occupy apes and all other animals for much of their daily lives. This chronic evolutionary pressure never goes away and nearly always (except during the worst famines) exerts itself in small increments. She who gets the best nutrition will have the most energy to compete for more food. He who feeds himself best will have the most energy to fight for mates. So the best place to start building a better theory of bipedalism is to consider whether standing upright is ever an advantage when it comes to finding and eating food.

I began my current study of chimpanzees without any special interest in their walking upright; after all, chimpanzees don't do it very often except in circuses. In fact, in the years that I spent watching chimps in Gombe National Park in Tanzania in the 1990s, I can't recall ever seeing chimpanzees stand or walk upright. I have visited Mahale National Park in Tanzania, where the Japanese primatologist Toshisada Nishida directs a famed study of chimpanzees. Nishida had initiated his work several years after Jane Goodall, who was only about sixty miles away, and had confirmed many of her pioneering discoveries. Mahale is a landscape of rugged, lushly forested hills, with many small rushing streams. While following the chimpanzees in the company of the Japanese research team one day, I ended up alone for a moment with one of the chimps as it approached the edge of a stream. Chimpanzees don't particularly like water; they're fascinated by it but avoid get-

Chimpanzees occasionally walk upright. This one is stepping out of a rushing stream in Tanzania.

By pulling down small branches, a chimpanzee standing on or near the ground gains access to new sources of fruit.

ting in too deep. This chimpanzee, a big male, waded through. As he reached the other side a few yards away, with water swirling around his legs, he stood up. Looking for all the world like a swimmer emerging from the surf, he reached up, plucked some leaves, stuffed them in his mouth, and then, grasping a plant for support, stepped out of the streambed into the forest.

This is bipedalism. But this behavior is rare, something an observer might see a few times a year in the life of a wild chimpanzee. But as I soon found out in my next study of wild chimpanzees, bipedalism doesn't have to be rare, and it doesn't have to take place on the ground. One morning in Bwindi Impenetrable National Park, Uganda, I watched a female chimpanzee in my study group stuff herself with figs at the top of a towering tree. With her baby clinging precariously to her chest, the mother reached above her head while standing on three limbs on a huge branch. Suddenly, the mother stood upright and used her long arms to reach several feet above her head to grab the figs. For thirty seconds she stood on two legs, her muscular feet anchoring her on the branch and her hands steadying her above. Her baby rode along, bobbing up and down as her mom plucked handfuls of figs and shoved them into her mouth. Other chimpanzees in the tree were standing upright too, although never for more than a minute or so and always using secure handholds on branches. Sometimes an ape would reach too far from the branch for a fig, and as its body leaned into space, it would reach for other branches, ending up almost horizontal, with each of its four limbs grasping a different tree branch.

Meanwhile, on the ground below, other chimpanzees were stuffing themselves with fallen fruit. One spied a nearby sapling that looked to be full of ripe fruit. He knuckle-walked to it, then reached up to pluck its fruit. First, he reached up and pulled the lowest branches down to his level, so he could sit on the ground and eat to his content. But some of the lowest branches were just out of his reach, so he stood upright and, grasping the branch with one hand, plucked its fruits with the other. He occupied himself this way for a few minutes, until his contented food grunts — "uh-uh-uh" — proved too good an advertisement and some of his group mates strolled over to check out his bounty.

We don't need great imagination to consider how this scene might have played out early in human history. Same forest, similar ape. But when the ape spots a small tree growing a few feet farther away from the last, he awkwardly takes three steps across the forest floor while standing upright, then sits down on his haunches to continue feeding at the new site.

It may not seem as important as a hominid's walking many miles in a perfectly erect posture, but it is bipedalism nonetheless. Whether on the ground or in treetops, these behaviors, repeated millions of times over a million years, could produce a selective advantage for the lineage of apes that was able to exploit food resources that their ancestors could not. Such an advantage could be the result of natural selection for whatever anatomical tweaking might have improved the ape's ability to stay upright longer and with more stability. This fortunate ape would have perpetuated its genes and therefore the genes behind the slow, steady transition to upright posture.

The primatologist Kevin Hunt of Indiana University, who did extensive research on the use of the bipedal posture by wild chimpanzees in Mahale National Park, Tanzania, in the 1990s, found that nearly one fifth of his observation hours contained incidents of bipedalism, both on the ground and in trees. Four of five times that his chimpanzees stood bipedally were in the context of feeding and foraging. Hunt sees both terrestrial and arboreal two-legged walking by wild chimpanzees as a clue to what early humans were doing when they foraged along the edges of the forest mosaics where they lived. He used his observations on chimpanzees' upright foraging to revive and revise the arm-hanging theory of Sir Arthur Keith. Whereas Keith's emphasis was on arm hanging while moving through trees, Hunt stresses the importance of using arm support to stand on tree branches.

Hunt was not the first scholar to suggest that the roots of bipedalism are visible in the behavior of modern apes. In the 1970s Russell Tuttle had argued that even gibbons, the long-armed swingers of Asian rain forest canopies, give a hint of bipedalism's origins when they walk upright along high tree limbs. He observed that such infrequent kinds of bipedalism could well have been precursors of what later took place when an ancestral ape began spending more and more time on the ground. Hunt and Tuttle had revived and revised the old arm-hanging theory of Sir Arthur Keith, except that Hunt proposed arm-hanging rather than arm-swinging. Bipedalism was used in the context of standing on tree branches with arm support, not swinging through trees. Although chimpanzees are inefficient upright walkers compared with humans, bipedalism practiced in concert with other postures may have been highly effective. And the primatologists Clifford Jolly and Richard Wrangham had independently complemented Tuttle's arboreal hypothesis during the 1980s, arguing that once on the ground, the sort of short-distance shuffling that we see in modern chimpanzees and a few other primates might be similar to what the protohominid was doing before real upright walking emerged.

The anatomist Michael Rose made the important point that although chimpanzees may seem inept when they walk bipedally, we must ask "inept compared with what?" Certainly, they are inefficient walkers compared with humans. But remember that at each stage of the evolutionary process, natural selection molded anatomy in the context in which it was used. When bipedalism is used in concert with many other postures in a foraging context, the sort of bipedal walking or shuffling that an apelike hominid practiced may have been just fine. Rather than view the earliest biped as a half-formed version of modern bipeds, we need to view it as a creature highly adapted to its own form of walking.

The benefit of viewing the origins of a two-legged stance this way is that it requires no deus ex machina, no special event, and no clean break from a four-legged past. The earliest bipeds did not have to shuffle through an adaptive black hole in order to "get to" better bipedal posture. For a knuckle-walker that is occasionally bipedal to become, over thousands of generations, gradually more bipedal requires no sudden or dramatic change in either behavior or anatomy. This effectively cuts the legs off models like Rodman and McHenry's — the energetic efficiency theory that butted heads with Karen Steudel's analysis of the energy equation of walking. Rodman and McHenry, along with Lovejoy, Stern, Susman, and the others, operate in a black-and-white world where one is either a good biped or a poor biped. None of that makes any sense when viewed in the light of what modern apes do. Early hominids were good at the kind of bipedalism that they practiced all along; changes in bipedalism occurred because of ever changing environmental pressures.

But why did the impetus for change occur at all? The environment in East Africa at five to six million years ago was changing, but savannas were not simply replacing forests, as we have seen. Instead, the type of forest was changing, from dense and closed to more open, as rainfall declined and the degree of seasonality rose. The distribution of preferred foods in those forests also changed as a result. Fruit trees that grew in closely packed stands became more dispersed; thickets that once grew like carpets now became patchwork quilts. To move about in these patchwork arrangements, even for just a few miles, hominids increased the frequency of their bipedal walking.

If bipedalism evolved as an adaptation perfectly attuned to a particular way of getting food in a forest, we can account for the many types of bipedalism that paleontologists are beginning to find in the fossils. The 1999 discovery of *Kenyanthropus* — the 3.5-million-year-old biped that is more modern than *Australopithecus afarensis* from the same time period — shows that bipedal species took a variety of evolutionary pathways. This diversity existed for a good reason. Gorillas and chimpanzees occupy the same forests in Africa today, using their food resources differently and avoiding head-to-head competition. Five million years ago many kinds of early humans were also feeding on different foods or feeding on the same foods but in different ways. The small differences in how much time they spent feeding on the ground or in trees translated into anatomical differences and would have allowed two or three types of protohominids to live in the same patch of forest.

In this discussion I have not tried to claim a single factor that led to bipedalism, nor have I invoked a simple, linear, one-step transition. Like so many other events in evolution, it just didn't happen that way. We have no reason to assume that the cause of the first steps was directly related to the cause of later improvements. Each step was a tweaked version of the step before it. Natural selection works on the package in hand, and once that package changed, the starting place for all future human evolution changed with it. Convincing the scientific community of this is difficult, because scientists are as gullible about good stories as anybody else. Tweaking is not as appealing a story as food carrying, concealing ovulation, or many others. But unlike the others, it actually matches the real world of ape behavior and the real way that evolution by natural selection works.

So far, I have left out one enormously important piece of the puzzle. While the events that I describe led to our ancestors' first steps, I haven't yet considered why, once the earliest humans were walking upright, bipedalism continued to become ever more efficient. Why did an apelike biped, capable of walking short distances at slow speeds for purposes of feeding, evolve into a marathon walker capable of almost limitless high-efficiency standing and walking?

7 THE SEARCH FOR MEAT

THEY BROKE CAMP at dawn, walking into the forest single file. Although they were little more than five feet tall, each man carried an enormous wooden bow that nearly dwarfed him. Some of the women came too; most would follow later after breaking camp. They followed no path that you or I could identify but walked with determination toward the rising sun. Their pace could best be described as a semi-jog; they moved with amazing alacrity through the tangle of roots and plants that grew on the forest floor.

By midmorning the men had traveled five miles, stopping only briefly to catch an armadillo. As the day progressed, they also found and collected honey, some grubs, palm fruits, and an agouti, a small South American rodent. But only as afternoon approached dusk did they find what they were looking for. A distant crashing in the forest canopy meant the presence of monkeys — capuchins, which the humans relished. The hunting party stopped, and one man whistled softly, imitating the sound of a young capuchin in distress. They waited patiently as he whistled again and again. Finally, a lone monkey appeared in the treetops about a hundred yards away, then another and another. The entire capuchin group moved toward the sound that the monkeys presumed meant a baby in trouble.

As the monkeys grew near, one hunter raised his bow and

fired an arrow, striking a monkey broadside and knocking it from the tree; it landed nearly at the feet of the archer. The other monkeys ignored the death of their comrade and continued to search for their lost offspring. One by one the hunters' arrows brought down the capuchins. Sometimes a primate was injured by the arrow but fell to the ground very much alive, where hunters caught and killed it with their hands. In all, the hunting party killed seven monkeys before moving off.

The men signaled their location to the women and children who were following far behind them, and as darkness fell in the forest, the group got together again, building shelters for the night and roasting monkey meat on the campfire. After a long day of walking — about fifteen miles — the band of hunter-gatherers retired, well aware that the next day would bring more of the same.

The central theory about the roots of human behavior and ecology is that, after the evolutionary road forked, separating humans and apes, our own ancestors increasingly began to eat meat. Meat eating is widely thought to have been the key factor in promoting the ballooning brain size that occurred in our more recent ancestors. But the notion was born in controversy.

In 1968 the eminent physical anthropologist Sherwood Washburn and his student Chet Lancaster published a paper, "The Evolution of Hunting," in a scholarly tome called *Man the Hunter*. Attempting to explain how and why the human brain had ballooned in size and complexity during the past several million years, the anthropologists wrote, "Our intellect, interests, emotions, and basic social life . . . are evolutionary products of the success of the hunting adaptation." They argued that big-game hunting was central to human origins. They also pointed out that in many traditional societies, men tend to be hunters, and women tend to be food gatherers.

Washburn and Lancaster thereby placed men in the all-important job of bringing home the bacon. Successful hunting requires communication and coordination, so becoming a hunter required intelligence and the ability to communicate. The authors also equated the male desire to hunt with the equally deep male desire to go to war. They suggested that men might naturally hold the glamorous role of providing meat, while women "merely" stayed home, gathered roots and tubers, and did the cooking. Theirs was the first of a flood of influential papers claiming the natural supremacy of males in the art of the hunt.

Man the Hunter was based on some valid logic. We know of many cases in which genetic linkages, called pleiotropic effects, cause one sex to exhibit traits that were produced by natural selection in the other sex. Male nipples are an example. But some anthropologists perceived a sex bias in this idea. The anthropologists Nancy Tanner and Adrienne Zihlmann of the University of California, Santa Cruz, pointed out in 1976 that in some hunter-gatherer societies, women rather than men obtain the vast majority of animal protein. Men might kill a giraffe once a year and then boast about it around the fire at night for a year until they managed to kill another giraffe.

The backlash to *Man the Hunter* was vituperative and long lasting. It led anthropologists and other scientists to re-evaluate the way in which sex biases might have hindered their interpretations of primate societies. Observers had usually tended to focus more on the behavior of male rather than female animals, because males are often bolder and therefore easier to watch. The backlash even led to a recognition that the role of women in early human societies had long been ignored. Stone tools made for butchering carcasses remain in the fossil record, whereas tools fashioned of wood and used for gathering by women would not have been preserved.

All these changes in thinking contributed to the untimely demise of the notion that meat eating was at the core of human existence. Only in the 1990s did meat eating once again take its place as the catalyst for much of human evolution.

In nature, meat comes in two forms: alive and dead. To catch living prey one must be a skilled predator, which means having either biological weaponry — claws and teeth — or crafted weapons such as spears. The earliest hominids had neither. How, then, could they have gotten meat? We can safely assume from studies of their teeth and their habitats that our earliest forerunners were mainly plant eaters, with fruit the likely food of choice. Our ancestors would have eaten meat in whatever form it was available. Both modern chimpanzees, themselves avid consumers of meat, and modern hunter-gatherer people eat animal protein, everything from insects to elephants. Catching an occasional rat, rabbit, piglet, fawn, or other small animal would have been a daily activity for hominids even without weapons.

Meat was widely available in the East African forests and grasslands of three to five million years ago; herds of hoofed game lived there, as they do today. But to catch these you need much better technology than little protohominids possessed. If you wait for a zebra or wildebeest to die on its own, or perhaps to be killed and partially eaten by a lion, you may have a large meal without the hassle of killing it yourself. The problem is that feasting on a big zebra carcass lying out in the African sun, appealing though it may be, is dangerous. Lions may be lying in wait in the nearest thicket, ready to ambush anyone cheeky enough to steal scraps from their kill. And even without the danger from big carnivores, a carcass in the African sun begins to rot soon after death and within a few days will be rendered inedible. So hoping to stumble across a kill is not going to be a reliable strategy for regularly finding meat if you're a meat-hungry evolving protohominid.

This dilemma — meat, meat everywhere but not a whole lot to eat — has perplexed human evolutionary biologists for decades. It has led to fierce arguments about whether early humans were hunters or scavengers and at what stage of our ancestry meat eating became important. After the demise of man-the-hunter theories in the 1970s, a new way of seeing meat eating came to the fore. Lewis Binford, an archaeologist and long-time firebrand for the "new archaeology," had long advocated reconstructions of the past that involved rigorous hypothesis testing by using the present day to understand the deep past. If you want to know how to determine whether an early hominid or a fossil hyena had chewed on a gazelle carcass, you need to know how a modern hyena eats a gazelle. Following Binford's example, archaeology students began to undertake behavioral studies of carnivores whose ancestors lived alongside early hominids. They also began to study how easy it is to misread fossil sites. Earlier archaeologists might have looked at a cave filled with the fossilized bones of hominids, antelope, and leopards and assumed that the hominids and leopards had both been predators that dragged the antelopes into the cave. A new generation of archaeologists looked at the same scene and realized that the leopards had dragged the hominids to the cave. Our ancestors were sometimes the hunted, not the hunters.

The rise of this more critical way of seeing the ancient world contributed, in the 1980s, to archaeologists' favoring scavenging, not hunting, as the way that our ancestors obtained meat. Many famous archaeological sites were re-examined and reinterpreted based on the new analyses, and what was once regarded as evidence for big-game hunting by early humans was reinterpreted as scavenging. Hunting as an impetus in human evolution fell by the wayside and was replaced by the notion that our ancestors were runty little creatures on a perpetual treasure hunt, racing in to cut a morsel from a dead zebra unobserved by the carnivore that had made the kill, then scurrying away to eat in peace.

Why does it matter whether we were hunters or scavengers? The difference is enormous in reconstructing our ancestral way of life. To be a skilled efficient hunter without the natural weapons of a big carnivore takes years of practice, learning, and careful observation of one's elders. It may also be critical to learn to coordinate your actions with those of your fellow hunters, because cooperation almost always improves your chances of killing a dangerous animal without getting yourself killed. These learning factors suggest a strong role for the social group and a survival value in being a member of a close-knit social group. They also imply that social intelligence is critical to becoming a good hunter, as is spending key formative years growing up with elders who can teach you the ropes.

Scavenging requires a very different set of skills. Group cooperation may not be as important as individual attention to the environment. Animals that make their living by scavenging usually have advanced skills in finding carcasses on a vast landscape. Smell is the sense of choice, but vision is also important. Modern foraging people home in on carcasses by watching for vultures circling from miles away. Did early hominids have these skills? Becoming a good scavenger involves a learning curve but perhaps one that is not as long or intensive as that to become a hunter, simply because your quarry doesn't move or defend itself. The only tactical problem is finding the darn thing on the vast Serengeti. Most researchers who have studied the practicality of subsisting on a diet of dead animals, with an eye toward understanding what early humans may have done, have concluded that scavenging is not a feasible way to make a living. Carcasses are widely scattered, decompose or are quickly eaten by other scavengers, and are often only seasonally available, during birthing and migration seasons and such.

CLUES FROM CHIMPANZEES

Meat eating is an unparalleled way to get calories, protein, and especially fat and is therefore the central human dietary adaptation.

Diet, and the foraging required to get food, are the most important influences on the behavior changes that herald the anatomical changes in the shift to standing upright. In order to understand how walking and meat eating are related, we draw key lessons from two populations, hunting and gathering societies, and meat eating but nonhuman primates. Of the latter, there are only a few, because primate diets are almost exclusively plant foods.

The renewed interest in meat eating came about partly because of new information about the importance of meat eating among chimpanzees. Of the apes, only the chimpanzee is an avid habitual meat eater. Although only a small percentage of the chimpanzee diet is meat, these apes are highly efficient predators that hunt cooperatively, share meat ritualistically, and value meat far more than plant foods, which are easier to obtain. Bonobos catch monkeys sometimes but don't eat them; instead, they carry them around like dolls or toys, inflicting nothing more than benign neglect and some rough handling before setting them free again. These apes eat small antelopes, but their overall level of meat consumption seems to be a fraction of that routinely seen in chimpanzees.

Chimpanzees rarely go out searching for meat; they almost always hunt after stumbling across small mammals while foraging for fruit. Nearly all chimpanzees in equatorial Africa live in forest habitats; even those that live in regions of grassland or along river courses spend nearly their entire life in the patches of forest that exist there. The prey most often taken by wild chimpanzees is colobus monkeys. These noisy treetop dwellers tend to stand and fight rather than flee, and if the attacking party of male chimpanzees is large enough, at least one kill is almost certain.

Wild piglets and antelope fawns are also frequent fare. The chimps find them hiding on the forest floor. The apes grab them and run while the parent pig or antelope makes a futile fuss in the thickets. Some of my most fearful moments during chimpanzee research have come when confronted at close range by an angry mother bush pig, as she waves her tusks and flashes her beady eyes at me. But the chimps steal piglets nonetheless, and what follows is a melee of captors, beggars, and onlookers jostling for a place at the table.

The hunting of monkeys, however, has captured the most attention from both scientists and the public. Unlike the frenzied spontaneity of a pig or bushbuck hunt, the capture of monkeys requires perfect execution and sometimes a good deal of foresight. As a foraging party of chimps treads the hills of an African forest on its daily rounds, it comes across many fruit trees, innumerable leafy shrubs and patches of herby plants, and an occasional group of monkeys. Red colobus are tree-loving monkeys that weigh up to twenty pounds and live in groups of more than fifty members. When attacked by a predator, the male colobus in the group band together to do courageous battle with the intruder. When chimpanzees pass by a tree holding a group of colobus, they pause, craning their necks into the canopy in hope of spying a mother colobus carrying a small baby.

If the apes decide to hunt, the result is often a gruesome slaughter. The male chimpanzees — hunting is largely a male pastime — climb the tree, approaching the cluster of anxious monkeys from below. Mothers gather up their babies, and male colobus attempt to form a protective screen between their mates and offspring on one side and the attackers on the other. I have seen male chimpanzees stymied by a knot of defending colobus that cluster at the first large fork in a tree's branches, keeping the chimps out of the tree crown the way a metal ring prevents a squirrel from reaching the top of a bird feeder. In circumstances less favorable to the colobus, however, the chimpanzees manage eventually to break the ranks of the monkeys and gain close access to their intended quarry, the females and their young.

Much of what happens next depends on the circumstances:

how many male chimpanzees are attacking and who the individual hunters are, how many male colobus are defending, how tall the trees are in which the attack takes place, and whether there are adjacent trees to which the monkeys can make their escape. Some male chimpanzees are avid, bold hunters. Frodo, the current alpha male in the Kasakela chimpanzee community in Gombe National Park, Tanzania, is one such eager predator. His presence enhances not only the odds that a hunt will take place but also the chances that it will succeed. And Frodo is a ruthlessly efficient hunter. During one five-year period in the 1990s, he alone killed roughly 10 percent of all the red colobus monkeys living within the territory of his community. In what I like to call the "Great Chimp Theory of History," one male hunter such as Frodo could have decimated the colobus monkey population.

The kill of a monkey is at once fascinating and awful to watch. Through years of research at Gombe the thrill of new observations and insights never quite overcame my revulsion at seeing monkeys I had watched grow up die at the hands of the chimpanzees. At Gombe male chimpanzees use one another to make a kill, without necessarily coordinating their actions. One male will chase a colobus mother carrying her infant to the tip of a branch, and as she leaps to another branch to escape, a second male is waiting there, either to attack her or to pluck the infant from her belly, allowing her to go free. This is not cooperation; the captor will not necessarily share the meat of his reward with those who helped him during the chase.

But whether the chimps share seems to depend on the chimpanzee society one is watching. Some chimps do share, to pay back those who helped in the kill or as a form of nepotism. At Gombe sharing is all about opportunism, kinship, and favoritism. Family members share with each other and snub nonrelatives, males share with other males and females with whom they need to solidify or build alliance networks, and males share with females who are in estrus and potential mates. Every member of the hunting party sits waiting, sometimes for hours, to receive a share of the kill, and many go away hungry. Nevertheless, a hunt can be the focal point of a day in the life of a chimpanzee, and the commotion often draws community members to the scene from far and wide.

This sort of behavior cannot help but intrigue anthropologists who specialize in reconstructing how our earliest ancestors behaved, what they ate, and how their diet may have influenced our own evolution. The first hard evidence of meat eating — simple, egg-shaped stone tools with one sharp edge — dates from 2.5 million years ago. Before this time most of us have no doubt that early humans were making and using some sort of tool — perhaps made of sticks or unmodified stones or even bamboo stakes — but nothing remains preserved in the archaeological record of such ancient technology. And the evidence from chimpanzees is that primitive hominids both hunted and scavenged, displaying the same cultural diversity from site to site that traditional human societies show today. Chimpanzee societies living a few hundred miles apart have different tool cultures and different meat-eating practices; we have no reason to think that the earliest humans would have been any different.

The rigid herbivory of gorillas stands in stark contrast to the avid omnivory of chimpanzees. We once believed that gorillas and chimpanzees were strikingly different in their behavior in many ways, but in the 1990s we began to realize that their differences are not quite so stark. We used to believe that gorillas were slow, sedentary leaf eaters who browsed in mountain meadows like primate versions of cattle. These bold dichotomies existed largely because the earliest, and still the most detailed, study of wild gorillas was that begun in the 1970s by Dian Fossey in the Virunga volcanoes of east-central Africa.

The tiny remnant population of gorillas in the Virungas is

not like other gorillas. These three hundred apes live in a mountain habitat far colder and more inhospitable than any other place in Africa where gorillas are found. Their diet of tough, fibrous leafy stuff — wild celery, thistles, and the like — is born not of desire but lack of options. Few fruit trees live at the elevations of 10,000 to 12,000 feet that the gorillas inhabit, so a carbohydrate-rich diet is out of the question. The Virunga gorillas merely hang on in a marginal habitat by subsisting on stuff that all other gorillas in Africa would turn up their noses at.

Across the rest of equatorial Africa, the 80,000 or so gorillas that make up the vast majority of gorilladom make it clear that the Virunga population is a sore thumb, ecologically speaking. Lowland gorillas live in vast tropical rain forests and in scrubby patches of forest just outside villages. Wherever they live, they live for fruit, falling back on a leafy high-fiber diet only when lean times force them to. In this they are unlike chimpanzees, who do not have the luxury of using poor-quality forest foods as a backup. When ripe fruit is scarce, chimpanzees must break up into smaller and smaller parties and travel ever farther to find the odd tree still bearing something fruity. But the more fruit that the lowland gorillas eat, the farther they travel as well, sometimes immensely farther than their montane brethren.

Gorilla social behavior is also quite different than we used to think. The word *harem* embodies the lifestyle of a silverback gorilla; he is the picture of power and domineering strength, outweighing "his" females by a hundred kilos. Such an imposing bulk, together with his macho chest-beating demeanor, led early researchers in the Virungas to consider silverbacks the lords of their domain and the sole leaders of their social group. It isn't quite that simple, however. Gorilla groups, we now know, often have multiple silverbacks, making the old harem concept inappropriate. Among eastern African gorilla populations in particular, including Fossey's famed apes and those in my study site in the

Impenetrable Forest, a large percentage of gorilla groups include two, three, and occasionally five or six silverback males. A group of gorillas with several silverbacks can be an exciting place, for researchers who have to be wary of getting in the middle of squabbling four-hundred-pound giants and, no doubt, for female and juvenile gorillas as well.

So contrary to the long-held conventional wisdom, gorillas in most of Africa are far from the inactive, celery-munching, harem-group apes that they are still often made out to be. Given their strong parallels of diet, activity patterns, and mating systems with chimpanzees', we would expect that the two apes might compete head to head for food and other resources when both live in the same forest. This certainly does happen, at least occasionally. In the Impenetrable Forest we have seen chimpanzees and gorillas arguing about possession of a fruit tree on rare occasions; it appears that in such contests, the smaller chimpanzees rule.

One key aspect of chimpanzee and gorilla ecology that differs greatly is meat eating. Chimpanzees crave meat and eat it at every opportunity, sometimes in large quantities. Gorillas, on the other hand, don't eat the meat of other mammals at all in the wild, even when it is easily caught. In zoos gorillas learn to relish beef and eggs and other fatty foods that we like to eat ourselves. Decades of information about captive diets has taught us something very important about fatty foods and gorillas — it kills them. American zoos no longer feed their gorillas any animal fats, having learned by tragic experience that even small rations of beef, eggs, and whole milk products lead to high rates of heart disease in gorillas, especially among silverbacks. This stands in marked contrast to chimpanzees, which may live to a ripe old age on diets of Kentucky Fried Chicken and french fries in the care of well-intentioned owners.

We are obsessed with the evils of fat and cholesterol consumption; check any health-oriented publication and try to sort through the often conflicting data on the bad and healthful influences of HDL and LDL cholesterol. But we tend to forget that, as a species, we are extraordinarily immune to the harmful effects of animal fat. For millions of years our ancestors did not have to worry about eating more milligrams of cholesterol than their middle-aged arteries could handle. They rarely survived to middle age, dying as a result of drought, disease, and predators long before. And the foods that we obsess about overconsuming today are the same foods that were in such short supply in eons past that they were worth their weight in, well, fat. You gorged on bird eggs in the spring when the birds were nesting, then didn't see a yolk until the following year. If you were skilled and lucky enough to bring down a zebra or giraffe, you and your family and allies ate it until you could barely walk. Hunger, not heart disease, was the imminent health risk.

This difference between the gorilla's meat-sensitive herbivory and our long history of carnivory almost certainly involves genetic mutations to enable meat eating. As my colleague at the University of Southern California, the gerontologist Caleb Finch, and I have argued, meat does a body good, at least when consumed in the limited amounts that conditions allowed during nearly all our evolutionary history. Just as chimpanzees have a higher tolerance of dietary cholesterol and fat than their gorilla cousins do, we have a dramatically greater ability to withstand the harmful effects of a fatty diet while reaping its enormous caloric and nutritional benefits. Somewhere in the early hominid lineage, a mutation probably arose that conferred an advantage on an early human who could stuff him- or herself with meat.

The presence of "meat-adapted" genes in our genome is more than just speculation. Biomedical researchers have found correlations between invulnerability to cholesterol and vulnerability to other, seemingly unrelated diseases, such as Alzheimer's. They suggest that natural selection may have built high tolerance for animal fats into the human genome, while the genes that confer the protection may be linked to Alzheimer's. In prehistory people simply didn't live long enough for Alzheimer's to be an important health risk. Besides, most Alzheimer's victims are beyond the age of reproduction, placing the disease almost out of reach of the power of natural selection to eradicate it. Ironically, our ability to live on a meaty diet in prehistory may have been genetically connected to the rise of a debilitating disease that affects millions of people today.

MEAT AND POTATOES?

Not everyone today believes that meat eating was all-important in becoming human. The Harvard anthropologist Richard Wrangham and his colleagues recently claimed that eating potato-like tubers was what led to the expansion of the human brain. They find convincing, if highly circumstantial, evidence for tuber eating and tuber cooking in the fossil record of two million years ago. We know that humans grew rapidly in stature and brain size during that time. Wrangham's group posits that early humans had a "eureka moment," not unlike that of the monolith-gazing ape-humans in *2001: A Space Odyssey*. Lightning-set fires are common on the African savanna, and some of the plants in the region have adapted to the risk of lightning fires by storing their energy underground, out of harm's way.

The Harvard researchers see a turning point in human origins in the cooking of these foods at a central place. There, females began to bond with particular males to obtain protection from those who would try to steal the cherished tubers. In forming such alliances, females would have extended their period of sexual availability, leading to less competition among males for mates. In turn, this would have lessened the extreme male-female size differences. And the rise of later hominids such as

Homo erectus would have heralded a new mating system, monogamy.

This appealing story would account for our pair-bonding tendency, so unusual among primates. The premise that cooking tubers ever occurred is, however, a speculation. It could well have happened, but the evidence for meat eating in this same period of human evolution is overwhelming. Without some hard evidence that tubers were part of the human diet two million years ago, we are left with mashed potatoes.

WALKING TALL IN TWO STEPS

While parallels exist between chimpanzee and early human hunting, there are also major differences. Chimpanzees never use weapons during a hunt or kill very large prey (thirty pounds is about their maximum). Most important, chimpanzees search for fruit and hunt only when the opportunity presents itself, the opposite of what hunter-gatherers do.

If meat is such a valued source of protein, fat, and calories, why don't chimpanzees wake up every morning and go looking for it? The answer lies in the difference between knuckle-walking and upright walking. A biped can afford to walk many miles across a landscape, gathering nuts, honey, and other foods as he goes in search of meat. At the end of the day, if the biped hasn't found any game to hunt, he has fed himself with the fruits, insects, and other foods gathered in the course of his marathon hike.

A chimpanzee, on the other hand, cannot spend an entire day searching for meat, because knuckle-walking is an inefficient way of getting around. Both chimpanzees and people are omnivores, making many decisions each day about what to eat and how much time and energy to spend getting it. But hunter-gatherers travel far and wide and do so efficiently. Chimpanzees knuckle-walk long distances in search of fruit, but to search relentlessly for game meat and fail to find it would be disastrous for the ape. Walking upright changes this equation.

To go from a four-legged ape to a two-legged human required two major steps. The first was quite modest. The ape who shuffled across the ground a bit between fruit trees gained an advantage over her or his mates. The ape began to change anatomically, because natural selection favored those that moved between trees efficiently. The partial biped could also use its new two-legged posture to reach up to low-hanging trees laden with fruit, much as chimpanzees do today. Some of this feeding would have happened in the trees too, just as I described the Bwindi chimpanzees doing.

Emerging hominids probably lived in a variety of habitats, and short-distance walking would not have been equally valuable in all settings. It might have been most important in forested areas where fruits grow in abundance near the ground, on small trees or bushes. Or it may have been most useful where forests were sparse, providing a strong incentive to change the way one moved between trees. Whichever the case, the earliest short-distance hominids would have begun to diversify as they invaded and occupied a variety of habitats and began to adapt to them. Within a few thousand generations a new panoply of creatures would have spread into diverse niches, practicing various sorts of bipedalism. Some of these we have discovered and studied; the fossilized remains of many others remain in the earth awaiting the light of day. And many more may have disappeared, leaving no trace of their tenure on the planet at all.

So by about seven or eight million years ago, the forests and grasslands of Africa were populated with apes and nascent hominids, a free-living zoo of numerous forms and structures that allowed two-legged walking. In all likelihood some bipeds adapted to dense forests where the need for walking was small, whereas other bipeds adapted to walking through lightly wooded country but always within easy escape to the trees. And still other bipeds adapted to sparsely forested plains. This last group grew in number as climate changes in East Africa about five to six million years ago produced ever widening expanses of grassland between the shrinking patches of forest. These early hominids began to walk ever farther afield. With each passing generation their anatomy began to reflect their new behavior. Thousands more generations passed, and efficient bipedalism took the place of shuffling or short-distance walking. The hominids that we identify in the fossil record today as *Australopithecus afarensis*, *A. anamensis*, and *Kenyanthropus* had arrived.

This set the scene for the second stage. As the hominids began to forage farther afield for fruit and other plant foods, their anatomy changed in just the way that theorists had long believed. The energy savings gained from efficient long-distance walking was what early hominids needed as their environment began to change from expanses of forest to greater expanses of grass. A rich array of hominids came to live in the mosaic of forest and grassland that occurred all across eastern Africa, and some species began to make extensive use of the more open country.

Once the level of walking efficiency allowed for longer-distance travel across more open country, hominids would have been able to do what their ape cousins could not: look for meat. As we've seen, scavenging or hunting requires lengthy scouring of the landscape for sources of game, and this burns much energy. But forage for meat they did, and over the generations their anatomies became more and more refined to bipedal walking. Contrary to most current theories of human origins, it is likely that even as the better bipeds were evolving, other types of bipeds clung to life in the forest, using their branch-pulling, between-tree-shuffling bipedalism to best advantage. *Ardipithecus ramidus*, Tim White's very primitive hominid, may well turn out to be one such early and ultimately unsuccessful variant of the "lesser biped" type of protohuman.

But small animals are just as available in forests as they are in grassland. Why did humans crave meat in their diet as they moved onto the savanna? Forest mammals tend to be solitary and highly secretive. But on the grassland hordes of hoofed mammals of all sizes and shapes live in great herds. They give birth, leaving offspring all over the landscape that become a readily available food source for any creature that knows when and where to look.

The most important source of meat on the grassland is one that early bipeds could only window-shop for. Killing a gazelle or zebra requires either weapons or group coordination. However, early hominids would have caught many small animals. And they scavenged. Early hominids added carcasses of big game to their diet of fruit, leaves, and small game, and this made all the difference. Large divisible sources of meat were the key for the evolution of intelligence, and efficient walking was the key for foraging for it. By three million years ago the whole equation of foraging energetics and diet had begun a fundamental shift; hominids became tool-wielding foragers who could drive lions off their kill. About two million years ago the brain began to show its true human dimensions. Not only that: Other changes, from body size to size differences between males and females, began to wreak alterations that reverberate to this day.

Meat eating, once established in the ever-more-efficient bipeds, played a key role in the feedback loop of diet, intelligence, and behavior. The anthropologist Katharine Milton of the University of California, Berkeley, has pointed out that the energy windfall involved in turning to a meatier diet is what allowed the apes to increase both body and brain size from chimplike to more human. She showed that until the quality of our ancestors' diet had risen as a result of becoming ever more carnivorous, our forebears had to spend too much time and energy searching for the right combinations of plant foods to evolve the levels of social complexity that exists in human, and presumably prehuman, societies. If Milton is right, meat was responsible for the second key event of human evolution after our shift to an upright posture: the emergence of intelligence. If walking upright enabled intensive foraging for meat, a whole cascade of other changes became possible.

Hillard Kaplan and his colleagues at the University of New Mexico have used evolutionary theory to explain the differences between the rearing of young people and of young apes. Kaplan argues that because meat is a rich package that everyone wants but is very difficult to catch, adults have to invest years in teaching their children to be the next generation of hunters. Learning to hunt takes time. Mother wolves and lions tutor their offspring for months before the youngsters can catch their own prey. Human children must apprentice for much longer.

Kaplan believes that this accounts for the dramatically greater period of parental care that is invested in children. The payoff is great: Skilled hunters will provide for themselves and their families through good years and lean ones for decades. So the protracted human life history, in which adulthood is delayed for years and years compared with other primates', may be a function of the need to invest parental time and energy in producing a productive member of a meat-avid culture. And as hunter-gatherer children mature into adults, their daily return of meat rises steadily. The end result is that people in traditional societies capture far more game than a chimpanzee could ever hope to.

Our brain size grew steadily for millions of years and then rather suddenly mushroomed. Some anthropologists believe that a crucial step occurred just under two million years ago, as the apelike australopithecines began to surrender Africa to the earliest humans. Other recent studies, however, place the date of the great brain explosion much later, at only about 250,000 to 300,000 years ago. The paleoanthropologist Bernard Wood of George Washington University argues persuasively that until this recent expansion, all earlier increases in brain size were simply following proportional increases in overall body size. He points out that only in the most recent human species, the early stages of modern humankind, did the expansion of brain volume accelerate beyond what an individual's body size would predict. If his conclusion holds up, then we must explain not just why brain size increased in human evolution but what happened 300,000 years ago that set off this runaway growth.

8
BETTER BIPEDS

In 1984 the paleoanthropologists Alan Walker of Pennsylvania State University, Meave and Richard Leakey of the National Museums of Kenya, and the legendary Kenyan fossil hunter Kamoya Kimeu made a dramatic discovery in the dusty soil of Kenya. Eroding out of an ancient dried-up waterhole was the skeleton of a hominid. Not just a few smashed-up bone fragments, as is the norm for even the most exciting discovery of human fossils. A *skeleton.* As they removed piece after piece from the gully bank, Walker and the rest of the team slowly realized that they were uncovering the most complete early human remains that had ever been found, more complete than the vaunted Lucy discovery a decade earlier. Not only that: This specimen was obviously much more modern than any australopithecine. The skeleton was that of a teenage *Homo erectus* boy. The cause of death is unknown, perhaps disease or starvation. The young victim had fallen face down into the edge of a pond, where nearly all his remains were entombed for more than 1.5 million years. "Nariokotome boy," as he was dubbed, is of stunning importance, because the completeness of his skeleton, along with its antiquity, allowed Walker to make estimates of how the boy would have grown had he not met his untimely fate. Supposing that he grew in the same way as a modern teenager, the boy — estimated to be about thirteen at the time of death — would have been a gangly six-footer at adulthood, with modern limb proportions.

This was an astounding discovery — humanity had progressed from the fairly apelike early genus *Homo* at two million years ago to a strapping modern body type only a few tens of thousands of generations later. Something important must have happened during that short time span to drive our ancestors toward their modern form. Bernard Wood of George Washington University and Mark Collard of Washington State University have pointed out that all hominids more primitive than *Homo erectus*, including *Homo habilis*, should be pushed back into the genus *Australopithecus*. This is because *Homo habilis* and its close relatives seem so much more apelike than *Homo erectus* in light of the newest fossil evidence for humanity in the 1.5 to 2 million–year range. Although the brain of a *Homo erectus* like Nariokotome boy is several hundred cubic centimeters larger than that of an australopithecine, most of this increase is due to the larger body size of *erectus*. So the brain expansion that was so fundamental to modern people seems to have disconnected itself from body size increase and taken off exponentially with the advent of the most primitive forms of *Homo sapiens*.

Was there a change in the nature of bipedalism in this period of modernization? Various forms of upright walking had long before allowed emerging hominids to occupy a wide range of ecological niches, to be followed eons later by more efficient walking that allowed humans to exploit a crucial new food source — meat. But why and how did walking become more and more modern as generations passed? Was it because our ancestors became big-game hunters and, if so, when? Or perhaps they became highly skilled scavengers? If the australopithecines were more or less like upright chimpanzees, albeit with some fancier cultural innovations and better tools, the rise of the first really humanlike humans did not occur until much more recently. Or did it? In order to evaluate this we need to examine the forms that early humanity took between two million years ago and today.

The perception of early humans such as *Homo erectus*, and

A group of *Australopithecus afarensis* leaves its sleeping tree and heads off in search of food.

older forms of *Homo sapiens* such as the Neandertals, has swung like a pendulum in the history of fossil science. Scientists have at times made *Homo erectus* out to be nearly human, an ancient hunter-gatherer, and at other times have portrayed him as little more than a beetle-browed ape-man. The truth, no doubt, lies somewhere in the middle. *Homo erectus* lived in the Old World from about 1.8 million years ago to 300,000 years ago, first in Africa and thereafter all across the Old World from the forests of Indonesia to the plains of Spain.

Homo erectus was without question a major consumer of red meat and needed to have effective weapons to use during the hunt for game. The tool of choice of *Homo erectus* was the hand ax, a rough-hewn stone tool that most often resembled a teardrop. It had a tapered knifelike edge that had been made by flaking the stone; the uncut end may have sat in the palm of the tool user's hand. Hand axes are as abundant in some *erectus* archaeological sites as pens and paper would be in our own.

Other tools that *erectus* used were scrapers and cleavers. These instruments were the home appliances of their day; paleontologists find hand axes were used in nearly unchanged form for almost one million years. Think about that: A human with a brain not much smaller than yours used the same primitive rock tool without great advances in its design for 50,000 generations. In our world, in which electronic toys change every six months, such complacency is hard to imagine. It may be elegantly simple testimony to the value of conservatism in a harsh environment.

What did *Homo erectus* do with its stone axes? These were large, roughly sculpted devices, but we don't know whether they were used to kill antelope, butcher a carcass after a kill, murder enemies, or even to cut lumber, as we would with a circular saw. (Because you can't easily carve a long-bladed saw from rock, they may have carved a round-edged disk; as one section became dull, they would have gone farther around the edge to a sharper section.) Recently, a team of researchers led by Manuel Dominguez-Rodrigo of the University of Madrid found evidence of crystallized plant cells on the sharp edges of an ancient hand ax. The presence of the plant cells on the edge of a stone tool meant one thing to the researchers: *Homo erectus* was using its hand axes to sharpen wooden spears. If this interpretation is right, we have evidence of *erectus* as a skilled hunter, rather than just a scavenger.

We must not be fooled by *erectus*'s use of the same tool for a million years without significant improvements. Modern hunter-gatherer people use very simple tools and weapons to hunt big game quite efficiently. If *erectus* had the cognitive abilities that most researchers believe, *erectus* would have hunted cooperatively, using either spoken or gestural language to coordinate a hunt. And they would have been big-game hunters, needing to learn and remember details: the best places to find game, the migration patterns of the herds, and how to stalk and kill a wide variety of animals. Such hunters must swiftly butcher carcasses before the inevitable lions and other predators arrive. These people also likely lived in a complex society that both enabled and necessitated cooperation in all these tasks.

Pat Shipman and Alan Walker believe that *Homo erectus* was the first human ancestor capable of carrying out all these tasks. They point to the increase in brain volume, from the puny five hundred to six hundred cubic centimeters of the australopithecines' to nearly one thousand cubic centimeters in *erectus*. However, as we've seen, other anthropologists place the origins of both rapid brain expansion and big-game hunting much later, at as little as 250,000 years ago. And in the pendulum swing of human evolutionary research, *Homo erectus* was recently pushed backward again, when an international team of researchers published data on the pattern of tooth eruption in *Homo erectus*, indicating that its growth and development pattern more closely tracked that of apes than of modern people.

Despite all this evidence for how *Homo erectus* lived and obtained meat, we have little information about how its upright walking differed from that of its ancestors. Alan Walker's Nariokotome boy offers the best evidence. He was a highly efficient biped, not only a far cry from australopithecines but perhaps even more efficient a runner and walker than a modern marathoner. He had narrow hips and long-necked femurs, allowing him to maintain his balance over a vertically narrower plane than any modern runner can. Those narrow hips were made possible by the one thing that female *Homo erectus* did not have to do that a modern woman does: push a big-brained baby through her birth canal. But wouldn't those narrow hips make upright life precarious for *erectus*, both literally and figuratively? Walker's former student Carol Ward, now of the University of Missouri, and Bruce Latimer of the Cleveland Museum of Natural History undertook a study of the Nariokotome boy's spinal column. Nariokotome has the most complete vertebral column of any human fossil, giving them fertile ground for their investigation.

Nariokotome boy's backbone also had two intriguing differences from that of a modern human. He had an extra lumbar vertebral bone in his lower back. This seemed to reflect a retention of the primitive condition of the backbone of earlier hominids. The later reduction from six to five vertebrae may have had something to do with a changing diet in humans, from all plants to a meat-plant medley, that required a smaller proportion of the upper body to be devoted to the abdomen.

Second, the bones of the vertebral column had less surface area, and therefore less weight-bearing capacity, than those of our modern backbone. In those respects the Nariokotome boy was a slightly less-than-modern biped. Both conditions led Alan Walker and his colleagues to suspect that *Homo erectus* might not have been as efficient a biped as necessary to do the long-distance running, walking, and stalking that big-game hunting would have required.

Then Walker and colleagues turned from the backbone to the ears. The inner ear might seem an odd place to search for clues about human origins, but not to Fred Spoor of University College, London, who pioneered the study of the semicircular canal of the inner ear. This is the organ that helps to regulate balance. Although the organ is made of soft tissue, the vestibular system of the inner ear is encased in a hard-shelled labyrinth composed of coiled tissues, including the semicircular canals, that fossilizes. Using computer-generated scanning images, Spoor studied the bony labyrinth of a variety of primates and other animals. He was interested in seeing whether the shift from quadrupedalism to bipedalism was accompanied by changes in the body's system of maintaining equilibrium.

Spoor found that *Homo erectus* possessed a vestibular system more developed than any in the lineage of earlier hominids. This was exciting news for any researcher studying the roots of bipedal walking. It lent support to the claim that *erectus* was a more modern walker than any earlier form of human. It also suggested that those earlier forms of humans did not walk bipedally in the modern sense, because they had not evolved a well-developed vestibular system for doing so. It lends strong support to the contention that *Homo erectus* was, in many ways, the first modern biped.

No hard evidence exists to tell us whether *erectus* could talk, but it almost certainly possessed some advanced form of language, given its brain size and tool-using abilities. Whether the language was gestural — a sign language — or spoken is unknown. Walker's team believes that *erectus* may not have possessed speech, at least not in the modern form. As evidence, Walker and colleagues point to the vertebral column of the Nariokotome boy, which is smaller in cross section than that of a modern person. This means that the spinal cord does not function as efficiently in motor control as it does in a modern person. This was puzzling. Normally, a mammal needs rich enervation in order to run its motor control system, and a creature like *erectus*, which was an efficient biped with dexterous hands for tool use, should have possessed an advanced system too. But it didn't. Relying on some anatomical work by the biologist Ann MacLarnon, Walker and colleagues surmised that Nariokotome boy had fewer neural pathways in his spinal cord than a modern human and therefore may have possessed only a primitive form of language.

Recall that one of the many ripple effects of walking upright was that breathing and striding became disconnected. This may have been a major evolutionary advance, one that allowed for the innovation of speech, because our ancestors could for the first time modulate their breathing rate to make speech sounds. And perhaps, Walker suggested, the reason that modern humans have so much more nerve enervation than *Homo erectus* is their need to enhance the control of breathing that makes speech possible. This would explain why *erectus* displays so much less enervation of the thorax. It needed fine control of its arms and legs, but it didn't have the same need for its diaphragm and chest that we do.

We don't know much about the origins of language. Because the body tissues involved in speech don't fossilize, we can only guess which of our progenitors first began to speak. Early hominids, including *Homo erectus*, may have used their hands to communicate with a sophisticated sign language, much as great apes can be taught to do when reared that way.

HUMANITY WRIT LARGE — THE NEANDERTALS

The label *Neandertal* took on a pejorative connotation decades ago; it implied boorishness at best and stupidity at worst. But the depiction of the Neandertals as brutish cavemen makes no more sense than if a future race of humanity were to depict us as such today. Discovered by miners in the Neander Valley near Düsseldorf, Germany, the first specimen came to light in the 1850s. No one knew what to make of it. Most experts considered it evidence of barbarians in the ancestry of modern Europeans. Ignorance about the role of Neandertals became compounded as time passed.

In 1908 another Neandertal was found in southwestern France, at a site known as La Chapelle-aux-Saints. The skeleton was nearly complete and lay in a fetal position. Fragments of animal bones and flint tools were strewn around the body. The find was turned over to the French anatomist Marcel Boule, who conducted a lengthy analysis of the skeleton and published an even lengthier book about his work. Boule concluded that the creature had been a brute in life, of low intelligence and perhaps not even fully upright in posture. Frankly, his analysis was terrible. He mistook arthritis in a crippled Neandertal man for a naturally hunched-over, primitive posture and on that basis characterized the creature as a dull-witted brute. In addition, Boule and his colleagues no doubt had difficulty avoiding the mind-set of the time, that early humans were by definition extremely primitive. The irony of the mistake was that an elderly, infirm individual could have survived only in a society that nurtured loved ones, a sure sign of intellectual sophistication.

The Neandertals were a form of nearly modern people that lived in the Old World from sometime more than 100,000 (and perhaps as much as 300,000) years ago to 30,000 years ago. The most recent evidence, based on the extraction of DNA from three Neandertals from different sites, is that the species was quite different genetically from fully modern people who were emerging at the same time. Scientists had to study the Neandertal fossil carefully to distinguish them from modern people. The differences are subtle — a longer flatter skull, a more prominent face, arching brow and jutting jaw, and a big-boned physique that suggests a modern Scandinavian. The Neandertal brain was actually larger on average than that of modern people, owing to the larger skull.

The discovery in the 1960s of Shanidar cave in northern Iraq, where Neandertals had been buried with what appears to be ceremonial care, complete with skin dyes and flowers, set science on the road to humanizing Neandertals. But the pendulum swings for this species as it does for *Homo erectus.* One of the world's foremost authorities on Neandertals, Erik Trinkaus of Washington University in St. Louis, Missouri, analyzed Neandertal pelvic bones and concluded that Neandertal women carried their babies for nearly a year rather than the modern nine months. But Trinkaus also had a keen interest in the implications that the Neandertal body had for the way it walked. He saw evidence in the Neandertal's robust lower body of poor locomotor efficiency. Trinkaus's findings were bolstered by the work of the archaeologist Lewis Binford, who in the 1970s and 1980s re-examined many fossil sites and showed that what appeared to be evidence of Neandertal hunting was often actually a natural process. The cut marks Binford found on hominid bones had been made by carnivores' teeth, giving the lie to the notion that brave hunting hominids had killed their prey. More likely, said Binford, Neandertals had merely absconded with carcasses killed by other carnivores. And deposits of bones were as likely to represent the natural actions of water and wind as the hand of man.

Both Trinkaus and Binford claimed that Neandertals were inefficient, disorganized hunters that must have wandered the landscape, aimlessly looking for big game and using their brawn to compensate for limited intelligence. According to the researchers, Neandertals lacked not only language but also the cognitive skills needed to coordinate a hunt.

The idea is almost certainly dead wrong. Apes, wolves, and bumblebees alike forage efficiently across the land using a memory for landmarks that even a primitive hominid would have possessed on a far greater scale. To suggest that Neandertals used only their powerful legs to stride ever farther in the hope of bumping into a herd of deer is a fanciful view. Research by many other anthropologists has shown otherwise. John Speth of the University of Michigan sees Neandertals as formidable hunters that sought prey in bands and were able to kill even the biggest and most dangerous animals of the Ice Age in Europe and Asia. He has gleaned this evidence from the Kebara cave in Israel, one of the world's most important Neandertal sites. Here, Neandertals lived seasonally in caves and hunted antelope skillfully. They did not kill only young and sick prey. The hunters were concerned about the health of their prey and tried to take healthy adults, even though the sick or juveniles would have been easier to catch. This selectivity is a distinctly human trait, not one that "dull-witted" Neandertals should have possessed. The humans cooked their kills in the caves, dumping their leftovers next to the cave wall.

In another important study the archaeologist Mary Stiner of the University of Arizona found that Neandertals in the Mediterranean killed and ate healthy adult prey, suggesting again that hunting rather than scavenging was the norm. Only humans routinely hunt healthy adult game, which tend to be a better source of fat than a young or infirm animal. Stiner's work showed that ancient European people, both anatomically modern and Neandertal, scavenged and hunted for a living and did so with great efficiency.

Are Neandertals and modern people one and the same, the former only racial variants of us? Most fossil experts reject that view. But some point out that by using the most widely applied criteria for anatomical modernity, aboriginal Australians might not be considered modern humans because of the bold bony brow ridge above their eyes. This just points out again the arbitrariness of assigning names to various forms of humans.

THE EMIGRANT TRAIL

They walked for days. Sometimes they gathered nuts and plants. On other occasions they hunted or found meat. Walking was second nature, and their long legs carried them into territories that no modern humans had ever passed. They walked across wide grassy valleys, over rugged hills, and through meandering rivers. The landscape changed gradually from tropical grassland to more seasonally cold, wooded environments and then became arid. They sometimes stopped in a place for months, even years. By the time they reached the Mediterranean, generations were present that had not been born when the journey started. As they went, they adapted their diet, clothes, and culture to fit the environment in which they found themselves. They engaged in that most human of all behaviors — they coped.

At some point in recent history the ability to walk efficiently for great distances allowed our ancestors to migrate from Africa to other parts of the world. Eventually, one generation of modern humans reached the easternmost tip of the Mediterranean Sea and entered the Middle East. From there their descendants spread into Eurasia, Europe, and points east. People did not suddenly arise one day and decide that the time had come to leave Africa. More likely, migrations happened in successive waves, and even then you might not have detected mass movement if you had been there to witness it. Moving just a few miles a year, the world's human population quickly inhabited areas beyond its equatorial cradle.

Exactly when this migration began is a point of intense debate. Until the late 1980s we believed that *Homo erectus* first left Africa about one million years ago. The thinking was that only after the evolution of highly efficient bipedal walking, combined with the rise of hand tools, were humans equipped to make the long hike required to reach other parts of the world. These earliest emigrants could not have been our own *Homo sapiens.* All evidence pointed to the origination of anatomically modern people 35,000 to 40,000 years ago. Then new evidence from both South Africa and the Middle East demonstrated that the appearance of fully modern-looking humans occurred roughly 100,000 years ago. Then, in June 2003, the announcement of the discovery of a modern human skeleton from Ethiopia with an estimated age of 160,000 pushed the date of origin back even further.

Even given the growing antiquity of our species, we were not the first humans to leave the African continent. Although several research teams have reported human fossils from Asia — from Indonesia to China to Eurasia — the date of one million years ago for our emigration from Africa fell with a thud with the discovery in 1999 of human remains at Dmanisi, a village about fifty miles southwest of the Georgian capital of Tbilisi. There, the ruins of a medieval castle sit at the fork of two rivers. The castle stands atop one of Eurasia's most important archaeological sites and the most important location in the region for discoveries of early humans.

The excavation at Dmanisi has uncovered the remains of at least four early humans, plus thousands of stone tools and the remains of thousands of ancient deer, giraffes, and other animals that the community presumably ate. The age of the Dmanisi fossils causes the most excitement and controversy. At 1.7 million years old the hominids at Dmanisi are by far the oldest human remains ever found on the European landmass and may be the earliest group of people known to have left Africa.

The Dmanisi humans are similar to the version of *Homo erectus* that first appeared in Africa 1.6 million years ago — Nariokotome boy. They possessed a very small brain, about midway in size between that of Lucy and that of a modern person and smaller than that of other, later *Homo erectus* specimens found in Europe. They were shorter and more slender than Nariokotome too. Their tools were not the hand axes of later *Homo erectus* but more roughly hewn chopping stones.

The Dmanisi people represent a perfect missing link between their ancestors from Africa and their descendants who lived in Europe and the Far East. Fossil experts hotly debate whether the Dmanisi version of *Homo erectus* and earlier versions of *erectus* from Africa, Beijing, and Indonesia should be called the same species. This argument has dogged the study of fossil humans at other times during the past two centuries.

In April 2001 the species question was explored at a session at the annual meeting of the American Association of Physical Anthropologists (the meeting was called "Read Our Lips, No New Taxa"). The Dmanisi fossils are much more similar to early African than later Asian *Homo erectus*, but most experts frown on splitting the several varieties into formal species. They believe that, as time passed and the migration wave spread, populations simply became more isolated from one another and diverged through the accumulation of mutations. But at what point they became recognizable as a new species is a question that we cannot yet answer.

The implications of the find at Dmanisi are of great importance for understanding the evolution of our ancestors from apelike early bipeds to big-brained marathon walkers. We know that *Homo erectus,* the first human species that shared many of our modern physical features, appeared in East Africa just under two million years ago. Shortly thereafter they began to migrate from Africa into the Levant, the then-fertile swath of land connecting Africa to Eurasia. The Dmanisi hominids tell us that these people quickly reached Europe and took their place atop the food chain there.

This emigration, so rapid and far-reaching, must have been enabled by some new innovation in humankind. We don't yet know what that was. Scientists always assumed that when *Homo erectus* invented the hand ax, the species could exploit new resources, become better at meat eating, and spread successfully across the Old World. But the fossils of the Dmanisi people, with their primitive tools, show this not to be the case. Their brains were larger than those of earlier humans, but did some newly acquired intellect enable them to make their way across the vast distances and varied environments? The prevailing view is that larger body size, rather than any new cognitive skills, may have been the key to the rapid dispersal. Larger mammals tend to have larger home ranges, to travel farther every day and throughout the year in search of food. As human body size increased, beginning at about two million years ago, *Homo erectus* and its variants may have been better able to cover more ground in search of game and other foods. This wandering would have set the stage for the migratory tendencies later on. If this thinking is correct, we see once again that upright walking, not expanded brain size, was the key to humanity's conquest of the earth.

OUT OF AFRICA AND INTO THE ORIENT

In the summer of 2002, I was attending the final day of an international conference in Beijing. A sweltering heat wave had confined all the attendees to the frigid blast of the air-conditioned hotel lobby, where I overhead some people talking about their visit to the Peking man caves. As an anthropologist, I knew all about the famed fossils from old Peking; I taught students about them every fall in a course on human evolution. But I had no idea that the place that the fossil came from was reachable so easily from downtown Beijing. Neglecting the end of the conference, I jumped into a taxi and headed about twenty miles outside the city, to the little town of Zhoukoudian. The archaeological site is a former quarry carved into a hilltop, from which several well-preserved skulls of early humans were pulled in the late 1920s. Today Zhoukoudian is beautifully restored, a UNESCO World Cultural Heritage Site with a fine small museum, many caves, and few visitors.

Site 1 is a cave that arches up from the bowels of a ravine. A trail runs through the gully right up to the lower lip of the cave entrance. The rock overhang once was the roof of the dwelling of untold generations of *Homo erectus* families. Although the site was changed a great deal by the old mining operation, I could imagine standing there 400,000 years ago, as beetle-browed people traipsed in and out with the day's catch or with children in their arms. These people were the descendants, 60,000 generations later, of the humans who occupied Dmanisi. They had arrived and settled at the farthest reaches of eastern Asia.

The fossils of Zhoukoudian are all the more remarkable because of their storied history. In 1921 Swedish geologists found some animal bones and human teeth at the site, which had been locally famous among Chinese as a place to find "dragons' bones" for use in traditional medicine. The Chinese geologist Pei Wenshong of the geological survey of China developed the site as an archaeological dig, and in 1929 he discovered a human skull in one of the caves. It was a magnificent find, Pei reported to scientists in Beijing.

Among the scientists in Beijing who eagerly awaited a glimpse of the skull was Davidson Black, a Canadian physician and fossil scholar who was in China to teach anatomy to local medical students. Ironically, Black had trained in London with some of the same scientists who had recently rejected Raymond Dart's Taung child as an early human. Black immersed himself in the work at Zhoukoudian as soon as he saw the skull, which he recognized to be that of a very primitive human. After months of excavating the skull from surrounding rock, Black set off on a European tour, publicizing to the world that he had found the most important human fossil yet discovered.

Unlike Dart in Africa and Eugene Dubois in Indonesia, whose work was met with deep skepticism, Black had little trouble convincing the scientific community and the public that *Sinanthropus pekingensis* (Chinese man from Peking) was profoundly important for understanding human prehistory. (Peking man was later renamed *Homo erectus* when experts realized it was the same species as Dubois' Java man skull.) The reception for Black's findings may have been due partly to the strong racist bias at the time — that Asia was a plausible place for humanity to have originated, whereas black Africa was not. Chinese today, including some Chinese scientists, continue to maintain that China rather than Africa is the cradle of humanity, despite the ironclad evidence of Lucy and her kind.

Davidson Black became a famed scientist on the back of Peking man and continued the excavation in the caves at Zhoukoudian until his untimely death in 1934. If Black is forever linked with the discovery of Peking man, his successor at Zhoukoudian, Franz Weidenreich, is forever linked to its disappearance.

Weidenreich took over the study of Peking man. Under his

direction the excavation turned up more human skulls and associated bones in the mid-1930s. In all, remains from the caves represented at least forty individuals, including six fairly complete skulls. More than 100,000 stone tools and other human artifacts were recovered.

Then world events intervened. The Japanese had invaded China in 1933, and as their occupation reached Beijing and Zhoukoudian, they developed a keen interest in the Zhoukoudian skulls. Japanese soldiers harassed and ultimately killed some workers who were excavating the site, forcing closure of the dig. As tensions grew in the late 1930s, and many expatriate Americans fled China, Weidenreich fled also, taking as many measurements, photos, and plaster casts of the Zhoukoudian fossils as he could carry to continue the research abroad.

In late 1941 word came that the Japanese were planning to raid the laboratory in Beijing where the remaining fossils were stored. Chinese scientists approached the U.S. embassy for help. The fossils were carefully packed in wrapping paper and crated and placed on a train headed for the coast, where they were to be put aboard a ship bound for California. A squad of U.S. Marines accompanied the precious cargo.

But the plan fell apart. The shipment of Zhoukoudian fossils arrived at the port city of Qinhuangdao on December 7, 1941, the day the Japanese attacked Pearl Harbor and President Roosevelt declared war against Japan. The U.S. ship that was to receive the fossils never arrived, and the train carrying the fossils was ambushed by the Japanese Army as it approached the coast. The Japanese seized the train and its cargo and took the Marines prisoner. The fossils have never turned up. Thankfully, Weidenreich had the foresight to make those casts and photos. Although work continued at Zhoukoudian after the war, no significant human fossil fragment has ever been found.

The caves at Zhoukoudian were occupied from about

450,000 to 230,000 years ago, making their inhabitants relatively young versions of *Homo erectus*. Anatomically, the people looked much like Dubois' Java man, which lived about 400,000 years earlier. The only real changes in *Homo erectus* through all those millennia were cultural. The stone tools at Zhoukoudian became smaller and of better quality as the generations passed. People began to use flint and high-quality quartz, replacing the more rough-hewn quartz tools of the earliest occupation. This presumably made the people more effective hunters. Fire may also have been used in the cave, although researchers have recently questioned this contention.

In its upright large-bodied anatomy, *Homo erectus* remained much the same creature from its appearance nearly two million years ago to its extinction only 200,000 years ago, just 300,000 years after these people presumably started living in the farthest reaches of eastern Asia. Some scientists believe that in order to migrate to Indonesia and settle its myriad islands, *Homo erectus* must have built boats. If this speculation is true, at an early stage of history *erectus* would have had the human ability to adapt locomotor skills to every manner of environment — a powerful testament to the ability of people to move into and succeed in every conceivable niche.

DID EVE WALK OUT OF EDEN?

Homo erectus people, now highly practiced long-distance walkers, had traveled far. But then *Homo sapiens* came along and quickly became the sole human inheritor of the earth. We used to assume that *Homo erectus* simply evolved into *Homo sapiens* sometime between 250,000 and 35,000 years ago. We even have a category of fossil humans, called archaic *Homo sapiens*, that seem to represent a transition between primitive and modern humans about 250,000 years ago. The supposition had been that *Homo erectus* evolved first into Neandertals, which later evolved into fully modern people.

Then in the 1980s archaeologists excavating in both South Africa and in the Middle East found evidence of fully modern humans living about 100,000 years ago. At the mouth of the Klasies River near Capetown, South Africa, a cave mouth looks out over the southern tip of the African continent. While a climb to the cave is arduous today, 100,000 years ago the cave opened onto the rocky beach. Inside, researchers found the oldest known remains of modern humans, perhaps 115,000 years old. Two other sites in Africa also provide ample evidence of modern people. Farther north, a cave in Israel at a place called Skhul contains the remains of modern humans from about 110,000 years ago. Only a football field away sits another cave, Tabun, which contains Neandertal bones dating from the same general period. Likewise, another pair of caves found side by side in Israel, Qafzeh and Kebara, have modern human and Neandertal remains, respectively.

The contemporaneous occurrence of both Neandertals and moderns throws out any notion that modern people evolved directly from Neandertals and has convinced many anthropologists that we did not evolve in a direct line from *Homo erectus*. Two rival schools of thought about the role of migration and dispersal of people have emerged since the 1980s. The debate between them is as rancorous as any in science.

Milford Wolpoff, a biological anthropologist at the University of Michigan in Ann Arbor, and the archaeologist Alan Thorne of Australia National University became the leading proponents for the view that after *Homo erectus* emigrated from Africa to Europe and Asia, the species began to evolve into modern humans more or less simultaneously across that vast region. This approach has been dubbed the multiregional continuity theory. Wolpoff and Thorne propose that once *Homo erectus* settled the Old World, modern people arose independently from each *erectus* population. The researchers believe that limited migration occurred among the various emerging populations, so that all modern humans look more or less the same. The obvious racial differences have come about, Wolpoff and Thorne say, as a result of the isolation of people in China from those in Scandinavia, and so on.

For their smoking gun Wolpoff and Thorne point to a number of skulls of great antiquity in Asia and Australia. A series of skulls from Sangiran in Indonesia, for instance, displays telltale anatomical similarities to those of modern Indonesians, and the Chinese skulls from Zhoukoudian appear to be linked to the facial form of modern Chinese from Beijing. The reason for such similarity, the researchers maintain, is obvious. Chinese people today are directly descended from *Homo erectus*, which occupied the region 500,000 years ago.

The implications of Wolpoff and Thorne's ideas are, if correct, profound. The antiquity of *Homo erectus* means that if we *Homo sapiens* descended directly from them, living racial groups would have a long history of separation, perhaps more than a million years. This is plenty of time for genetic differences to accumulate. To some, it might even suggest that intellectual differences alleged to exist between racial groups had some scientific basis in the evolutionary history of the various geographic groups.

But this view did not sit well with many fossil experts, who view modern people as a new species, wholly different from any that came before. Many researchers were also uncomfortable with the idea of the antiquity of modern races that Wolpoff and Thorne had put forth. In 1988 Christopher Stringer and Peter Andrews of the British Natural History Museum proposed a new model for the rise of *Homo sapiens*. Trying to account for the seemingly rapid appearance of modern people all over the Old World, Stringer and Andrews claimed that we should view the transition from *erectus* to ourselves more as a species formation. In their view modern people arose from some ancestral stock in Africa, then rapidly emigrated, replacing every more primitive population of human as they went. Modern people, therefore, would not share an ancestry with Neandertals — the latter would be mere twigs on the human family tree, driven into extinction by our ancestors without leaving a genetic trace.

The perspective of Stringer and Andrews was radically different and had far-reaching implications for how we view ourselves as a biological species. They proposed that modern people arose quite late in the game, only in the past 150,000 years. This implies that modern human races only recently diverged and that any differences among them are probably biologically meaningless. Nowhere in the world, except in Africa, did modern people have any direct *Homo erectus* ancestry, and nowhere did Neandertals fit into our tree.

Having posed a provocative theory of recent migration that turned on the degree of relatedness among modern people, Stringer and Andrews and their growing army of advocates sought evidence. It came not from the dusty pits of archaeological digs but from the laboratories of biochemists and geneticists. The late Alan Wilson, a famous biochemist at the University of California, Berkeley, and the anthropology graduate student Rebecca Cann sought to determine whether the apparent similarities between ancient fossils and modern people living in the same place reflect a genetic line of descent from both.

Wilson and Cann intended to study the DNA of people from around the world to learn the true degree of relatedness among us. If the last shared ancestor of everyone living on Earth today was alive a million years ago, the DNA results would support Wolpoff and Thorne's continuity theory. If the last common ancestor lived much more recently, Stringer and Andrews' rapid replacement theory, which has often been dubbed "Out of Africa," would be proved. But there was a problem. Comparing the DNA of people from around the world wouldn't produce meaningful results, because changes in the DNA of a cell nucleus happen only very slowly as mutations in one population occur and accumulate, while different mutations occur in other populations. So Wilson and Cann creatively turned to another type of DNA that had been little used in the study of human origins up to that time.

Along with the DNA found in the nucleus of the cell, the mitochondrion, a small body found in the cell but outside the nucleus, possesses its own genetic code. Scientists believe this is an artifact from an ancient time when the mitochondrion was an entirely separate creature that eventually merged with cells but retained a bit of its autonomy. Mitochondrial DNA offers two unique properties. It mutates rapidly, so changes in its DNA sequence accumulate in far less time than DNA mutation in the nucleus. This allows scientists to compare differences in the genes of two people from, say, Nigeria and Denmark, whereas nuclear DNA from the pair would not reveal any distinctions at all. Also, mitochondrial DNA is inherited only through the maternal side; you possess a copy of your mother's DNA, your mother possesses your maternal grandmother's, and so on. In the absence of the messy introduction of genetic material from both sexes that occurs during sexual reproduction, scientists can more easily trace backward the inheritance pattern of someone's genetic makeup.

Wilson and Cann needed donors to provide the mitochondrial DNA, however. They received it from an unlikely source: placentas donated by women of various ethnic backgrounds who had given birth. With a rich supply of DNA samples the researchers set to work, extracting the mitochondrial DNA, analyzing the genetic differences among the women, and constructing a tree of their shared and separate ancestries.

The results were stunning. The entire six-billion-member population of the earth today is descended, the researchers claimed, from a single woman who lived in Africa a mere 140,000 years ago. Wilson and Cann went to great lengths to explain that this woman — inevitably dubbed Eve — was not the first woman of our species. She was just the one woman whose genes survived thousands of generations of mixing until the present. If this seems unlikely, consider an analogy involving the genes in a human society and the names in a telephone book. A telephone book is full of surnames, some rare and unusual and others that are extremely common. In a Hong Kong phone book you would find names like Lee and Wong predominating; in Mexico Sanchez and Gonzalez would be abundant; in Ireland, Kennedy and Moore would fill many pages. Why? Part of the reason is that some people have been more successful than others at perpetuating their genes and therefore their surnames. My maternal grandmother produced only daughters, and her daughters produced only sons. Thus my maternal grandmother's maiden name, Selg, disappeared when she died. In the same way her mitochondrial genetic makeup will disappear when her daughters die.

If the modern human races diverged as recently as 140,000 years ago, consequences for our view of human history are important. It means that racial differences are trivial at a genetic level and meaningless for those who would like to find biologically based intelligence differences between ethnic groups.

Wilson and Cann's study was certainly not unassailable. Wolpoff and Thorne attacked the mitochondrial study for obtaining its DNA samples from American women of various ethnic ancestries. Many African Americans have sizable percentages of European and Native American genes in their genetic makeup. The continuity advocates also claimed that the "clock" by which the geneticists measured the mutation rate of mitochondrial DNA has a large margin of error. If they were off only a bit, the date of supposed divergence at 140,000 years ago might be magnified to closer to one million years ago. In that case, Wilson and Cann would have observed only the original migration from Africa of *Homo erectus* and not a migration of the much more recent humans that the geneticists believed replaced *erectus*.

At this point the debate grew heated. Wolpoff and Thorne pointed out that if Wilson and Cann's interpretation was correct, modern people arose in Africa and then replaced every earlier form of human throughout the world (because otherwise the presence of earlier forms would be detected in the DNA studies). *Replaced*, in that case, is just a nice way of saying *killed*. While people have killed one another in the course of migrations throughout history, Wolpoff thought it unlikely for two species as similar as, for instance, Neandertals and modern people to live in the same place at the same time without at least sometimes fraternizing.

I find Wolpoff's argument persuasive. When the British sea captain James Cook sailed through Polynesia, he and his crew encountered small dark-skinned people whom, in the Eurocentric values of their day, they viewed as another species of human, inferior to Europeans. These same British sailors, however, had no qualms about having sexual relations with native women and fathering children with them, and British genes from those first encounters still exist in some Polynesian islander populations today. By the same logic, I imagine that at least occasional star-crossed pairings of modern humans and more primitive versions occurred.

Unlike scientific debates about human origins in decades past, this one did not find lines of white-coated lab scientists arrayed against field-hardened fossil experts. The original proponents of rapid replacement were, after all, Stringer and Andrews, both venerable fossil authorities. And some of rapid replacement's sharpest critics were geneticists. John Relethford of the State University of New York at Oneonta pointed out that although genetic evidence appears to have the imprimatur of "hard science," it is no less subject to error and misinterpretation than are fossils. He noted that a secondary wave of migration from Africa could give geneticists false results in their interpretation.

Genes, Relethford said, give us precious little reliable information about the timing of evolutionary events in human prehistory. Instead, the genetic evidence tells us much about the way that populations of people behaved in ancient times. For instance, we can trace the population explosion of humanity that occurred worldwide about 100,000 years ago simply by noting the sudden rise in diversity of the world gene pool at that time. That same evidence also suggests that the total number of early people alive before that population boom was tiny, perhaps tens of thousands in the entire world. But genes do not tell us whether the world population had been tiny just before the boom or was simply re-expanding after the sort of population crash that follows an epidemic or famine, such as has occurred many times in human history and, no doubt, prehistory.

Some anthropologists have tried to resolve the scientific head butting between the rapid replacement and multiregional advocates. They say that although early people spread out across the Old World, they probably engaged in extensive exchange of genes, not to mention cultural artifacts such as tools. These exchanges would have kept all people looking fairly uniform, both anatomically and genetically. Think of the multiregional model as a candelabra, with the departure of *Homo erectus* from Africa at the base and each candle representing Asian or European diasporas and simultaneous evolution toward modernity.

Now consider the same candelabra but with connecting brackets between the candles. This is the compromise model. We can call this partial continuity, and it makes much sense. Many rapid replacement advocates, however, do not accept partial continuity, perhaps because accepting some cross-species mating between ancient and more modern humans throws doubt on the validity of the rapid replacement presumption that all genes of earlier humans were replaced by those of moderns.

Nonetheless, the rapid replacement proponents have the upper hand at the moment. A 1995 study of the Y chromosome by a team led by Robert Dorit of Yale University has complemented the original mitochondrial DNA work. The Y chromosome is inherited through the paternal line and so makes a perfect test case for the maternally inherited mitochondrial DNA. The researchers found less genetic variation in the Y chromosome in humans than in any other species of primate. This suggested a very recent ancestry for *Homo sapiens*, because the genetic uniformity among many populations meant too little time had passed for variations to accumulate. At the same time another research team reported more variation in African versions of another chromosome than they found in the rest of the world combined. And a third study, of DNA extracted from a Neandertal bone, showed how genetically different Neandertals were from modern *Homo sapiens*.

Taken together, these three studies have offered strong support for the rapid replacement model. They show convincingly that, as a species, modern people are genetically uniform compared to other primates. But whether this uniformity necessarily means that we emerged very recently is still unclear. That there were successive waves of emigration, with geneticists seeing evidence of only the final one, remains a distinct possibility. Wolpoff and his fossil compatriots remain thoroughly unconvinced of the validity of the genetic approaches. Only time will tell whether we have descended from the conquerors of *Homo erectus* or are just the final wave of two-legged migrants who added their numbers to those already in the rest of the world.

THE MODERN BIPED: HOMO SAPIENS

By the time of the arrival of anatomically modern people, bipedal walking had been the law of the hominid land for more than 1.5 million years. People had not only been walking the long distances required in their migrations, they also used their long legs for finding meat as efficiently as modern hunting-and-gathering people do today.

The shift from pursuing to ambushing prey happened long

after modern people had become fully bipedal. It may have occurred when modern people began to occupy the whole of Eurasia 100,000 years ago. The archaeologist Lawrence Straus of the University of New Mexico thinks that this change occurred more recently, about 20,000 years ago, at least in Europe and southern Africa. European *Homo sapiens* ate the big mammals: bison, reindeer, red deer, and wild horse. *Homo sapiens* could have killed these potentially dangerous animals only through cooperation and much planning. These hunters, armed with spears, darts, bows and arrows, harpoons, and other projectiles, learned to ambush prey by chasing them into narrow canyons and to follow migratory herds in their annual wanderings. Hunting became strategic and well coordinated. And these modern walkers had the necessary locomotor efficiency and cognitive powers to follow, find, and stalk their prey.

As hunters acquired more technology through invention and contact with other cultures, they became progressively more efficient killers. Archaeologists have found evidence from 300,000 years ago, in the East African nation of Zambia, of a fossilized wooden club. From the southern coast of England comes a broken spear shaft made from a yew branch, also from 300,000 years ago. Almost 400,000 years ago, hunters in Germany were apparently using spears, judging from a recent discovery in Hannover of seven-foot wooden shafts found together with stone tools and the butchered remains of wild horses. Neandertals would have used wood to make digging sticks, hide scrapers, bottles and jugs, clothing, and of course building materials, as well as spears and clubs. In all probability people made tools of a wide variety of materials, not only wood but leaves and clay, few of which have survived.

Ancient hunters used all these tools in much the same way that modern hunter-gatherers use them today: to kill animals by ambushing or chasing them and to butcher their carcasses. But does being a good, efficient biped make you a better hunter? Being bipedal doesn't necessarily confer any great advantage while hunting — think of dinosaurs: Many bipedal forms were slow-moving herbivores. The benefit of bipedalism in recent hominids is at least threefold. It allows efficient marathon travel in search of game and other foods. It allows the agile pursuit of prey. And it allows people to carry tools, weapons, and children. This is a wonderful example of preadapation. The initial stages of walking upright were tentative, diverse, and suited mainly to modest feeding advantages. Only later did longer-distance walking become possible, let alone advantageous. And only much later than that did humans refine their newfound walking ability for marathoning. Walking evolved from quadrupedal apes to diverse protobipeds, with at least one lineage emerging as an accomplished long-distance biped. But at each stage of the trajectory, no end point was ever in sight, only an immediate genetic accommodation to a particular set of environmental circumstances.

THE MIGRANTS

You might still believe that people inherited the earth because of our brains. But I maintain it was all because of our ability to stand up and move from place to place in an efficient, upright manner. And when we consider how our ancestors colonized the earth, their walking ability was critical. All the early migrations of humans from Africa were strolls in the park compared to the voyages of modern people to settle the farthest reaches of the world. A group of people, perhaps as few as fifty, reached southwestern Asia about 100,000 years ago. From this tiny "founder" group the earliest agriculturalists emerged tens of thousands of years later. By 40,000 years ago people were making beautifully refined tools such as stone knife blades at a few sites in central Europe. By 30,000 years ago such tools were universal.

Something fundamental happened to humanity at that time. People expanded into the Arctic, the steppes of Russia, and other inhospitable places, carrying their ever-more-sophisticated technologies. Archaeologists believe that people began to live in larger settlements, eventually leading to a less nomadic lifestyle. And people began to depict their lives with paintings on cave walls throughout Europe. Symbolism and artistry either crept into the human psyche at that point, or for the first time humans were able to express themselves in ways that their descendants would someday discover. People also started to depict themselves, as in the enigmatic female-bodied Venus figurines that were carved all across Europe and may have played roles in fertility ceremonies, religious rituals, or just art.

When people first walked into the New World is unclear. Certainly they came by way of the Bering land bridge connecting Siberia and Alaska. Most archaeologists accept that by 13,000 years ago, as the Wisconsin ice sheet was in retreat, humans ventured forth and, perhaps traveling only a few miles a year — or perhaps journeying much farther — found the American continent. Some researchers believe an earlier wave of migrants had made it across as much as 25,000 years earlier, while the Ice Age was still going strong. Experts can say with certainty that 16,000 years later, people were living in Siberia very near Alaska, and that 3,000 years after that, ocean levels had dropped enough that the land bridge connecting the two continents had appeared. A trip across would not have been easy, for the land bridge was treeless and had a horrendous climate. By hunting whatever game animals occupied the vast area, the earliest emigrants to America struggled their way across during the ensuing millennia.

As people entered North America, they moved quickly, reaching Pennsylvania and other points east, for instance, by 12,000 years ago. Also, the first evidence of human settlement appears in South America at about this time. Climate changes were under way during this period — and as temperatures rose, more and more environments were available for rapid colonization. By 10,000 years ago the famed Clovis settlements of the eastern and central United States were booming, and shortly thereafter Clovis people, characterized by their spear points, were living all over the continent.

The native Americans were the ultimate extension of the six-million-year march to upright posture, because they were the first modern people to reach the point farthest from the origins of humankind in East Africa. Using boats, other people traveled throughout the Pacific, colonizing island groups as they went. But in the New World, it was the walkers who came and conquered, at a time when foot traffic was the only means of travel available.

9
SKY WALKERS

In a memorable scene in *Star Wars,* Luke Skywalker and his mentor, Obi-Wan Kenobi, walk into a sleazy tavern on the remote desert planet Tatooine. The bar, located in an intergalactic trading post, is frequented by extraterrestrial riffraff of all sorts. There are four-eyed dwarves, snake-headed men, froglike men, big furry dog-men, and fuzzy Dr. Seuss–like characters from other imagined worlds. Imagined by Hollywood, that is.

One anatomical feature shared by the aliens in the bar and in nearly all science fiction, from "Flash Gordon" to *Star Trek,* is that they tend to walk upright on two legs and to have dexterous hands with several, usually five, fingers. This is partly the result of the constraints imposed by central casting: Finding out-of-work actors with authentically extraworldly bodies is not easy. So every sci-fi character, at least before the age of computer-generated figures, bears a striking resemblance to a guy in a costume.

If we allow ourselves to think outside the terrestrial box, we can consider whether advanced life on other worlds might take far more bizarre forms. Having a certain number of digits per limb is a function of evolved developmental traits. Most lineages evolved five, while a few amphibians retain only four. Standing and walking upright are also the consequence of a long evolutionary history of becoming a biped. How critical is standing upright, so rare a feature of Earth's inhabitants, as a precondition for evolving sophisticated intelligence?

We are human in large part by virtue of our technology, so if our ancestors had not stood up and walked, tool use would not have reached its extraordinary levels. One look at Earth's other truly smart creatures tells us that having hands free to manipulate objects may be a necessary first step to evolving a human type of intellect. Chimpanzees are the most technologically skilled animals on Earth other than ourselves, but their posture severely limits their manual adeptness. I have watched chimps gathering stick tools to use as fishing probes at earthen termite mounds. They line up the sticks carefully between their lips to carry them, because they lack a free hand — knuckle-walking requires them to keep all four limbs on the ground. And dolphins, which roam the seas communicating with one another by using as sophisticated a set of signals as any nonhuman animal, have never developed the art of tool using. Their flippers, so wondrously honed for steering them through deep oceans, are useless for grasping anything. Indeed, only one form of tool use by dolphins has yet been discovered: They grasp natural sponges from the ocean floor, latch them to their noselike rostrum, and push them into the sea floor in search of food. After all, dolphins don't have a leg to stand on or a hand to grab with. Elephants use their nimble trunks to carry things and to pluck food from trees, but without digits their ability is stuck in low gear. Even among the lowly invertebrates, the species with the most developed "manual" dexterity, the octopus, is also the brainiest.

But back to the *Star Wars* tavern and the extraterrestrial question that it begs. If we someday receive a visit from smart extraterrestrials, will they look at all like us? Indeed, life in the rest of the universe may follow biological rules very similar to those on Earth. That is, Darwin's evolution by natural selection may be true of life everywhere. A rocky planet that orbits a star at the right distance should provide conditions hospitable to life. Add the right mixture of geology, water in some form, and temperatures that neither freeze nor bake critters, and some form of life may end up crawling all over the planet several hundred million years later. And the smart ones probably will walk upright.

This is more than pie-in-the-sky speculation. We have a perfect example of how the physical environment molds life in consistent ways right here on Earth. Forty million years ago Australia became an isolated continent because of rising ocean levels and continental drift. At the time primitive mammals living in that region became passengers on a continental ark and ended up as the dominant inhabitants of Australia. You know the rest of the story — these mammals found a landmass largely without competition and diversified into a vast array of forms that took advantage of every environmental niche. Left to their own evolutionary devices, Australian marsupials took the mammalian road less traveled, solving the reproduction problem by substituting a pouch for a uterus.

Australia offered a cornucopia of natural habitats not very different from those in the world in which the placental mammals were evolving. They had deserts, woodlands, rain forests, grassy plains, and seacoasts. Australia's marsupial mammals exploited all these, evolving into burrowers, climbers, runners, gliders, herbivores, carnivores, and anteaters, just as placentals did elsewhere. Marsupial mice, marsupial wolves, and marsupial lions all evolved, although the latter two are now extinct. A kangaroo, for all its leggy strangeness, is actually little different from a deer or antelope, except in the way it travels around. The many species of kangaroo, and wallabies large and small, occupy much the same role as grazers and browsers that hoofed mammals — the ungulates — occupy elsewhere. Natural selection exploited the grassland-woodland niche in Australia as it did in the rest of the world. It was just a novel down-under approach to the problem of getting around. Marsupial mammals, for all their bizarreness, speak to the pervasive influence of the natural environment and of natural selection in molding creatures. The striking convergence between marsupials and placentals tells us a great deal about what we should expect of extraterrestrial life. Given similar environmental conditions, we can expect that plant or plantlike communities should spring up. These should form the resource base for animal communities. Assuming that Darwinian rules are universal, animals not dramatically different from our own would evolve and diversify in Earthlike habitats. Otherworldly creatures would therefore not look all that otherworldly, or at least no more bizarre than some of Earth's forms. The platypus, with its nipples, duck's bill, and egg-laying reproductive technique, or the aye-aye, a practically toothless lemur that has a middle finger of obscene proportions, probably stretch the bounds of variation among the mammals as much as any E.T. would.

NASA is trying to find planets resembling Earth that are orbiting other stars in our galaxy. Sometime in the not-too-distant future we may find another world with conditions favorable to the evolution of intelligent life. What would they look like? There is no way to change the environment around you other than to tweak it by hand. Any intelligent life that we might encounter will have a body plan roughly similar to our own, or at least to that of a primate. The number of digits would be dependent on all sorts of developmental and genetic factors, but the presence of limbs and digits is almost a certainty.

Would an intelligent extraterrestrial need to walk upright? Bipedalism on Earth is so rare that it would be foolish to assume that it would arise again unless certain special conditions coincided. We might explore this by asking the same question about life on Earth. Was bipedalism a one-shot event, contingent on an alignment of myriad unlikely circumstances of environment, genes, and history? If so, it might not ever happen again if we were allowed to turn back humanity's clock by eight million years and play the evolution tape of apes and hominids over again. Or did the right ecological conditions make bipedalism nearly inevitable?

One factor is most important: The ancestor to a biped must have had a package of anatomical traits that natural selection can readily remold into the body of an upright walker. This involved extensive and sometimes labyrinthine changes in our own progenitors. The anatomical prerequisites were all there for the first hominids but perhaps not for the other mammalian groups that might have evolved uprightness. Those prerequisites included the grasping hand and opposable thumb. This most basic of primate adaptations had its origins more than sixty million years before the dawn of humanity. It didn't evolve for tool use or carrying but rather for moving about in trees and for catching the insects that formed part of the diet of the earliest primates. But the opposable thumb was a crucial holdover that natural selection could put to good use many eons later in a biped whose hands were free to learn to manufacture and use tools.

This also suggests the reason why bipedalism evolved in only one group, the primates (although we have seen that it may well have evolved multiple times in the earliest days of humanity). In addition to the greater energy efficiency of foraging bipedally, freeing the hands was a factor only in the primate lineage, because only primates had the intelligence to put their hands to good use in ways that aided their survival. The reward of bipedalism in the hominids may simply have been greater than for other mammals, and that is why it was pushed along.

Given a set of environmental parameters, many biologists believe that it is possible to make rough predictions about the general course of evolutionary change. Some anthropologists have written that primate mating systems have evolved along predictable lines. If we carefully examined all the relevant biological factors that might influence human evolution, they claim, we could retrospectively examine our immediate ancestors. But despite the best efforts of many human evolutionists, predicting future patterns in evolution, even retrospectively, is largely a guessing game.

Jack Stern and Randall Susman sum up the central issue of this book: "The more fossils that are found, the more we are surprised. Excitement builds for the discovery of specimens from the 4–6 million year old range . . . The challenge, we submit, may lie in our ability to identify this ancestor as a hominid."

Past generations of researchers worked on the assumption that bipedalism was the fundamental hallmark of humanity and that by finding the oldest biped they would find the deepest root of the human family tree. Now we know better. The oldest biped may simply be an upright ape. Of the several, or perhaps many, lines of early bipeds, one ancestor emerged whose descendants walk the earth today. The only legacy of the others is the bones filling dusty museum drawers and those still sitting in the ground waiting to be stumbled upon. Figuring out which of those dry bones represent upright apes, and which are the heart and soul of our humanity, is the challenge we face.

Cloaked in an oversized white suit and reflective glass helmet, an astronaut floats high above blue Earth. His gloved hands fumble with the tools he uses to do the most simple tasks. His booted feet are bolted to the robot arm of the space station, lest he drift helplessly away. The line between life and silent quick death is razor thin.

Two hundred miles below, a scuba diver explores a coral reef in deep blue waters. His feet are rubber finned, his body is wrapped in neoprene, and his head is bottled up in a mask and air hose that link him to the metal tank of oxygen on his back. Remove the tank, or cut the hose, and the diver is sentenced to certain death.

A few miles away, a rock climber is plastered against a naked cliff face. Her hands seek the smallest crevice for support, her feet are wedged into a crack below her. Without the metal pitons and ropes that support her, she might easily fall to an instant death, even if the fall were only a dozen meters.

We humans are hopeless misfits in all but a few of Earth's environments. Even on flat dry land, we can exist only within a narrow temperature range; above or below it we are doomed, unless we have technological aids. And yet, through the power of natural selection and a great deal of contingent history, we emerged from the billions of other kinds of animals in Earth's history to become the dominant species. Most of this history is hidden from us today, veiled by the mist of the ages. But what little of it we can see, and what more we can extrapolate, tell us that it all began with an ape that stood up and walked.

BIBLIOGRAPHY AND FURTHER READING

1. A FIRST STEP

Brunet, M., et al. 2002. A new hominid from the Upper Miocene of Chad, Central Africa. *Nature* 418:145–51.

Dart, R. 1953. The predatory transition from ape to man. *International Anthropological and Linguistic Review* 1:201–19.

———. 1959. *Adventures with the missing link.* New York: Harper.

Darwin, C. 1871. *The descent of man and selection in relation to sex.* London: J. Murray.

Elliot-Smith, G. E. 1923. The study of man. *Nature* 112:440–44.

Engels, F. 1896. The part played by labor in the transition from ape to man.

Gregory, W. K. 1930. The origin of man from a brachiating anthropoid stock. *Science* 71:645–50.

Haeckel, E. 1874. *Anthropogenie oder Entwickelungesgeschichte des Menshcen.* Leipzig: Englemann.

Huxley, T. H. 1863. *Evidence as to man's place in nature.* London: Williams and Norgate.

Keith, A. 1903. The extent to which the posterior segments of the body have been transmuted and suppressed in the evolution of man and allied primates. *Journal of Anatomy and Physiology* 37:18–40.

———. 1923. Man's posture: Its evolution and disorders. *British Medical Journal* 1:451–54, 499–502, 545–48, 587–90, 624–26, 669–72.

Landau, M. 1991. *Narratives of human evolution.* New Haven, Conn.: Yale University Press.

———. 1995. Paradise lost: The theme of terrestriality in human evolution. In J. S. Nelson, A. Megill, and D. N. McCloskey, eds.,

The rhetoric of the human sciences, pp. 111–24. Madison: University of Wisconsin Press.

Le Gros Clark, W. E. 1967. *Man-apes or ape-men?* New York: Holt, Rhinehart and Winston.

Osborn, H. F. 1928. The influence of bodily locomotion in separating man from the monkeys and apes. *Science Monthly* 26:385–99.

Schultz, A. H. 1953. The place of the gibbon among the primates. *Journal of the Royal Anthropological Society* 53:3–12.

Tuttle, R. H. 1974. Darwin's apes, dental apes, and the descent of man: Normal science in evolutionary anthropology. *Current Anthropology* 15:389–98.

———. 1975. Knuckle-walking and knuckle-walkers: A commentary on some recent perspectives on hominoid evolution. In R. H. Tuttle, ed., *Primate functional morphology and evolution*, pp. 203–9. The Hague: Mouton.

Washburn, S. L. 1963. Behavior and human evolution. In S. L. Washburn, ed., *Classification and human evolution*, pp. 190–203. Chicago: Aldine.

———. 1968. Speculation on the problem of man's coming to the ground. In B. Rothblatt, ed., *Changing perspectives on man*, pp. 191–206. Chicago: University of Chicago Press.

2. KNUCKLING UNDER

Conroy, G. C., and J. G. Fleagle. 1972. Locomotor behaviour in living and fossil pongids. *Nature* 237:103–4.

de Waal, F.B.M. 1987. Tension regulation and nonreproductive functions of sex in captive bonobos (*Pan paniscus*). *National Geographic Research Reports* 3:318–35.

de Waal, F.B.M., and F. Lanting. 1997. *Bonobo: The forgotten ape.* Berkeley: University of California Press.

Doran, D. M., and K. D. Hunt. 1994. Comparative locomotor behavior of chimpanzees and bonobos. In R. W. Wrangham, W. C. McGrew, F. B. M. de Waal, and P. G. Heltne, eds., *Chimpanzee cultures*, pp. 93–108. Cambridge, Mass.: Harvard University Press.

Fleagle, J. G. 1999. *Primate adaptation and evolution.* 2d ed. New York: Academic Press.

Fleagle, J. G., et al. 1981. Climbing: A biomechanical link with brachiation and with bipedalism. *Symposia of the Zoological Society of London* 48:359–75.

Gebo, D. L. 1996. Climbing, brachiation, and terrestrial quadrupedalism: Historical precursors of hominid bipedalism. *American Journal of Physical Anthropology* 101:55–92.

Gebo, D., et al. 1997. A hominoid genus from the Miocene of Uganda. *Science* 276:401–4.

Goodall, J. 1986. *The chimpanzees of Gombe: Patterns of behavior.* Cambridge, Mass.: Harvard University Press.

Hunt, K. D. 1992. Social rank and body size as determinants of positional behavior in *Pan troglodytes. Primates* 33:347–57.

Köhler, M., and S. Moyà-Solà. 1997. Ape-like or hominid-like? The positional behavior of *Oreopithecus bambolii* reconsidered. *Proceedings of the National Academy of Sciences* 94:11747–50.

Latimer, B. M., T. D. White, W. H. Kimbel, and D. C. Johanson. 1981. The pygmy chimpanzee is a not a living missing link in human evolution. *Journal of Human Evolution* 10:475–88.

Lewis, O. J. 1972. Evolution of the hominoid wrist. In R. H. Tuttle, ed., *Functional and evolutionary biology of the primates*, pp. 207–22. Chicago: Aldine-Atherton.

———. 1989. *Functional morphology of the evolving hand and foot.* Oxford: Clarendon.

Parish, A. R. 1994. Sex and food control in the "uncommon chimpanzee": How bonobo females overcome a phylogenetic legacy of male dominance. *Ethology and Sociobiology* 15:157–79.

Remis, M. 1995. Effects of body size and social context on the arboreal activities of lowland gorillas in the Central African Republic. *American Journal of Physical Anthropology* 97:413–33.

Stanford, C. B. 1998. *Chimpanzee and red colobus: The ecology of predator and prey.* Cambridge, Mass.: Harvard University Press.

———. 1998. The social behavior of chimpanzees and bonobos: Empirical evidence and shifting assumptions. *Current Anthropology* 39:399–420.

Videan, E., and W. C. McGrew. 2001. Are bonobos (*Pan paniscus*) really more bipedal than chimpanzees (*Pan troglodytes*)? *American Journal of Primatology* 54:233–39.

Walker, A., and M. Teaford. 1989. The hunt for *Proconsul*. *Scientific American* 260:76–82.

Ward, C. V. 1993. Torso morphology and locomotion in *Proconsul nyanzae*. *American Journal of Physical Anthropology* 92:291–328.

Wrangham, R. W., C. A. Chapman, A. P. Clark-Arcadi, and G. Isabirye-Basuta. 1996. Social ecology of Kanyawara chimpanzees: Implications for understanding the costs of great ape groups. In W. C. McGrew, L. F. Marchant, and T. Nishida, eds., *Great ape societies*, pp. 45–57. Cambridge: Cambridge University Press.

Zihlmann, A. L., J. E. Cronin, D. L. Cramer, and V. M. Sarich. 1978. Pygmy chimpanzee as a possible prototype for the common ancestor of humans, chimpanzees, and gorillas. *Nature* 275:744–46.

3. HEAVEN'S GAIT?

Abitbol, M. M. 1987. Obstetrics and posture in pelvic anatomy. *Journal of Human Evolution* 16:243–55.

———. 1996. *Birth and human evolution*. Westport, Conn.: Bergin and Garvey.

Aiello, L., and C. Dean. 1990. *Human evolutionary anatomy*. New York: Academic Press.

Carrier, D. R. 1984. The energetic paradox of human running and hominid evolution. *Current Anthropology* 25:483–95.

Chapman, G., N. G. Jablonski, and N. T. Cable. 1994. Physiology, thermoregulation, and bipedalism. *Journal of Human Evolution* 27:497–510.

Dainton, M., and G. A. Macho. 1999. Did knuckle-walking evolve twice? *Journal of Human Evolution* 36:171–94.

Falk, D. 1990. Brain evolution in *Homo:* The "radiator" theory. *Behavioral and Brain Sciences* 13:333–81.

Falk, D., and G. Conroy. 1983. The cranial venous system in *Australopithecus afarensis*. *Nature* 306:779–81.

Grant, R. B., and P. R. Grant. 1989. *Evolutionary dynamics of a natural population*. Princeton, N.J.: Princeton University Press.

Griffin, T. M., and R. Kram. 2000. Mechanics of penguin walking: Waddling walk does not explain expensive locomotion. *Nature* 408:929.

Hunt, K. D. 1994. The evolution of human bipedality: Ecology and functional morphology. *Journal of Human Evolution* 26:183–202.

Leonard, W. R., and M. L. Robertson. 1997. Rethinking the energetics of bipedality. *Current Anthropology* 38:304–9.

Lovejoy, C. O. 1988. The evolution of human walking. *Scientific American* 259:118–25.

Margaria, R., P. Cerretelli, P. Aghemo, and G. Sassi. 1963. Energy cost of running. *Journal of Applied Physiology* 18:367–70.

Pinshow, B., M. A. Fedak, and K. Schmidt-Nielsen. 1977. Terrestrial locomotion in penguins: It costs more to waddle. *Science* 195:592–94.

Richmond, B. G., and D. S. Strait. 2000. Evidence that humans evolved from a knuckle-walking ancestor. *Nature* 404:382–85.

Rodman, P. S., and H. M. McHenry. 1980. Bioenergetics and the origin of hominid bipedalism. *American Journal of Physical Anthropology* 52:103–6.

Rose, M. D. 1984. Food acquisition and the evolution of positional behavior: The case of bipedalism. In D. J. Chivers, B. A. Wood, and A. Bilsborough, eds., *Food acquisition and processing in primates*, pp. 509–24. New York: Plenum.

Rosenberg, K. R., and W. R. Trevathan. 1996. Bipedalism and human birth: The obstetrical dilemma revisited. *Evolutionary Anthropology* 4:161–68.

Ruff, C. B. 1995. Biomechanics of the hip and birth in early *Homo*. *American Journal of Physical Anthropology* 98:527–74.

Sanders, W. J. 1998. Comparative morphometric study of the australopithecine vertebral series Stw-H8/H41. *Journal of Human Evolution* 34:249–302.

Steudel, K. L. 1994. Locomotor energetics and hominid evolution. *Evolutionary Anthropology* 3:42–48.

———. 1996. Limb morphology, bipedal gait, and the energetics of hominid locomotion. *American Journal of Physical Anthropology* 99:345–55.

Tague, R. G., and C. O. Lovejoy. 1986. The obstetric pelvis of A.L. 288-1 (Lucy). *Journal of Human Evolution* 15:237–55.

Taylor, C. R., and V. J. Rowntree. 1973. Running on two or four legs: Which consumes more energy? *Science* 179:186–87.

Taylor, C. R., N. C. Heglund, and G.M.O. Maloiy. 1982. Energetics

and mechanics of terrestrial locomotion. *Journal of Experimental Biology* 97:1–21.
Trevathan, W. 1987. *Birth: An evolutionary perspective.* New York: Aldine de Gruyter.
Tuttle, R. H. 1974. Darwin's apes, dental apes, and the descent of man: Normal science in evolutionary anthropology. *Current Anthropology* 15:389–98.
———. 1975. Knuckle-walking and knuckle-walkers: A commentary on some recent perspectives on hominoid evolution. In R. H. Tuttle, ed., *Primate functional morphology and evolution*, pp. 203–9. The Hague: Mouton.
———. 1981. Evolution of hominid bipedalism and prehensile capabilities. *Philosophical Transactions of the Royal Society of London* 292 (series B):89–94.
Walker, A., and P. Shipman. 1996. *The wisdom of the bones.* New York: Alfred A. Knopf.
Wheeler, P. E. 1984. The evolution of bipedality and loss of functional body hair in hominids. *Journal of Human Evolution* 13:91–98.

4. THE EXTENDED FAMILY

Abourachid, A., and S. Renous. 2000. Bipedal locomotion in ratites (Paleonatiform): Examples of cursorial birds. *Ibis* 142:538–49.
Andrews, P. J. 1989. Palaeoecology of Laetoli. *Journal of Human Evolution* 18:173–81.
Berman, D. S., et al. 2000. Early Permian bipedal reptile. *Science* 290:969–72.
Brunet, M., et al. 2002. A new hominid from the Upper Miocene of Chad, Central Africa. *Nature* 418:145–51.
De Heinzelen, J., et al. 1999. Environment and behavior of 2.5-million-year-old Bouri hominids. *Science* 284:625–28.
Foley, R. A. 1991. How many hominid species should there be? *Journal of Human Evolution* 20:413–27.
Harrison, T. 1991. The implications of *Oreopithecus bambolii* for the origins of bipedalism. In B. Senut and Y. Coppens, eds., *Origine(s) de la bipédie chez les hominidés*, pp. 235–44. Paris: Centre National de la Recherche Scientifique.

Köhler, M., and S. Moyà-Solà. 1997. Ape-like or hominid-like? The positional behavior of *Oreopithecus bambolii* reconsidered. *Proceedings of the National Academy of Sciences* 94:11747–50.
Leakey, M. G., et al. 2001. New hominid genus from eastern Africa shows diverse middle Pliocene lineages. *Nature* 410:433–40.
McCollum, M. A. 1999. The robust australopithecine face: A morphogenetic perspective. *Science* 284:301–4.
Orians, G. H., and J. H. Heerwagen. 1992. Evolved responses to landscapes. In J. H. Barkow, L. Cosmides, and J. Tooby, eds., *The adapted mind*, pp. 555–79. New York: Oxford University Press.
Pickford, M., and B. Senut. 2001. "Millennium Ancestor," a six-million-year-old bipedal hominid from Kenya — Recent discoveries push back human origins by 1.5 million years. *South African Journal of Science* 97:2–22.
Reed, K. E. 1997. Early hominid evolution and ecological change through the African Plio-Pleistocene. *Journal of Human Evolution* 32:289–322.
Rook, L., et al. 1999. *Oreopithecus* was a bipedal ape after all: Evidence from the iliac cancellous architecture. *Proceedings of the National Academy of Science* 96:8795–99.
Sereno, P. C. 1999. The evolution of dinosaurs. *Science* 284:2137–46.
Spoor, F., B. Wood, and F. Zonneveld. 1994. Implications of early hominid labyrinthine morphology for evolution of human bipedal locomotion. *Nature* 369:645–48.
Suwa, G., et al. 1997. The first skull of *Australopithecus boisei*. *Nature* 389:489–92.
Tappen, M. 2001. Deconstructing the Serengeti. In C. B. Stanford and H. T. Bunn, eds., *Meat-eating and human evolution*, pp. 13–32. New York: Oxford University Press.
Tavaré, S., et al. 2002. Molecular and fossil estimates of primate divergence times: A reconciliation? *Nature* 416:726–29.
White, T. D., G. Suwa, and B. Asfaw. 1994. *Australopithecis ramidus*, a new species of early hominid from Aramis, Ethiopia. *Nature* 371:306–12.
WoldeGabriel, G., et al. 1994. Ecological and temporal placement of early Pliocene hominids at Aramis, Ethiopia. *Nature* 371:330–33.

5. EVERYBODY LOVES LUCY

Agnew, N., and M. Demas. 1998. Preserving the Laetoli footprints. *Scientific American* 262:47–55.

Berge, C. 1994. How did the australopithecines walk? A biomechanical study of the hip and thigh of *Australopithecus afarensis*. *Journal of Human Evolution* 26:259–73.

Berger, L. R., and P. V. Tobias. 1996. A chimpanzee-like tibia from Sterkfontein, South Africa, and its implications for the interpretation of bipedalism in *Australopithecus africanus*. *Journal of Human Evolution* 30:343–48.

Brunet, M., et al. 1995. The first australopithecine 2,500 kilometres west of the Rift Valley (Chad). *Nature* 378:273–75.

Clarke, R. J., and P. V. Tobias. 1995. Sterkfontein member 2 foot bones of the oldest South African hominid. *Science* 269:521–24.

Coffing, K. E. 1999. Paradigms and definitions in early hominid locomotion research. *American Journal of Physical Anthropology*, supplement 28:109–10.

Coppens, Y. 1994. East side story: The origin of humankind. *Scientific American* 270:62–79.

Fruth, B., and G. Hohmann. 1994. Nests: Living artefacts of recent apes? *Current Anthropology* 35:310–11.

Haüsler, M., and P. Schmid. 1995. Comparison of the pelves of Sts 14 and AL 288-1: Implications for birth and sexual dimorphism in australopithecines. *Journal of Human Evolution* 29:363–83.

Huw, R., et al. 1998. The mechanical effectiveness of erect and "bent-hip, bent-knee" bipedal walking in *Australopithecus afarensis*. *Journal of Human Evolution* 35:55–74.

Johanson, D. C., and M. Taieb. 1976. Plio-Pleistocene hominid discoveries in Hadar, Ethiopia. *Nature* 260:293–97.

Johanson, D. C., and T. D. White. 1979. A systematic assessment of early African hominids. *Science* 202:321–30.

Johanson, D. C., et al. 1982. Morphology of the Pliocene partial hominid skeleton (AL 288-1) from the Hadar Formation, Ethiopia. *American Journal of Physical Anthropology* 57:403–52.

Joulian, F. 1994. Culture and material culture in chimpanzees and early hominids. In J. J. Roder, B. Thierry, J. R. Anderson, and N. Herrenschmidt, eds., *Proceedings of the Fourteenth Congress of the International Primatological Society*, pp. 397–404. Strasbourg, France: Université Louis Pasteur.

Jungers, W. L. 1982. Lucy's limbs: Skeletal allometry and locomotion in *Australopithecus afarensis*. *Nature* 297:676–78.

Kingston, J. D., B. D. Marino, and A. Hill. 1994. Isotopic evidence for Neogene hominid paleoenvironments in the Kenya rift valley. *Science* 264:955–59.

Kramer, P. A., and G. G. Eck. 2000. Locomotor energetics and leg length in hominid bipedality. *Journal of Human Evolution* 38:651–66.

Latimer, B. 1991. Locomotor adaptations in *Australopithecus afarensis*: The issue of arboreality. In Y. Coppens and B. Senut, eds., *Origine(s) de la bipédie chez les hominidés*, pp. 169–76. Paris: Centre National de la Recherche Scientifique.

Leakey, M. D., and J. M. Harris. 1987. *Laetoli: A Pliocene site in northern Tanzania*. New York: Clarendon.

Lovejoy, C. O. 1978. A biomechanical review of the locomotor diversity of early hominids. In C. J. Jolly, ed., *Early hominids of Africa*, pp. 403–29. New York: St. Martin's.

———. 1988. The evolution of human walking. *Scientific American* 259:118–25.

———. 1993. Modeling human origins: Are we sexy because we're smart, or smart because we're sexy? In D. T. Rasmussen, ed., *Origins and evolution of humans and humanness*, pp. 1–28. New York: Jones and Bartlett.

Lovejoy, C. O., K. G. Heiple, and A. H. Burstein. 1973. The gait of *Australopithecus*. *American Journal of Physical Anthropology* 38:357–80.

MacLatchy, L. M. 1996. Another look at the australopithecine hip. *Journal of Human Evolution* 31:455–76.

McHenry, H. M. 1991. Sexual dimorphism in *Australopithecus afarensis*. *Journal of Human Evolution* 20:21–32.

———. 1994. Behavioral ecological implications of early hominid body size. *Journal of Human Evolution* 27:77–87.

McHenry, H. M., and L. R. Berger. 1998. Body proportions in *Australopithecus afarensis* and *A. africanus* and the origin of the genus *Homo*. *Journal of Human Evolution* 35:1–22.

Rak, Y. 1991. Lucy's pelvic anatomy: Its role in bipedal gait. *Journal of Human Evolution* 20:283–90.

Sept, J. 1998. Shadows on a changing landscape: Comparing nesting patterns of hominids and chimpanzees since their last common ancestor. *American Journal of Primatology* 46:85–101.

Stanford, C. B., and J. S. Allen. 1991. On strategic storytelling: Current models of human behavioral evolution. *Current Anthropology* 32:58–61.

Stern, J. T. 1975. Before bipedality. *Yearbook of Physical Anthropology* 19:59–68.

———. 2000. Climbing to the top: A personal memoir of *Australopithecus afarensis*. *Evolutionary Anthropology* 9:113–33.

Stern, J. T., and R. L. Susman. 1983. The locomotor anatomy of *Australopithecus afarensis*. *American Journal of Physical Anthropology* 60:279–317.

Susman, R. L., and J. T. Stern. 1991. Locomotor behavior of early hominids: Epistemology and fossil evidence. In B. Senut and Y. Coppens, eds., *Origine(s) de la bipédie chez les hominidés*, pp. 121–32. Paris: Centre National de la Recherche Scientifique.

Susman, R. L., J. T. Stern, and W. L. Jungers. 1984. Arboreality and bipedality in the Hadar hominids. *Folia Primatologica* 43:113–56.

Tague, R. G., and C. O. Lovejoy. 1998. AL288-1 — Lucy or Lucifer: Gender confusion in the Pliocene. *Journal of Human Evolution* 35:75–94.

Tardieu, C. 1979. Aspects biomèchanique de l'articulation du genou chez les primates. *Bulletin de la Société Anatomique du Paris* 4:66–86.

Tuttle, R. H. 1981. Evolution of hominid bipedalism and prehensile capabilities. *Philosophical Transactions of the Royal Society of London* 292 (series B):89–94.

———. 1987. Kinesiological inferences and evolutionary implications from Laetoli bipedal trails G-1, G-2/3, and A. In M. D. Leakey and J. M. Harris, eds., *Laetoli: A Pliocene Site in northern Tanzania*, pp. 503–23. Oxford: Clarendon.

Tuttle, R. H., D. M. Webb, and M. Baksh. 1991. Laetoli toes and *Australopithecus afarensis*. *Human Evolution* 6:193–200.

White, T. D., and G. Suwa. 1987. Hominid footprints at Laetoli: Facts and interpretations. *American Journal of Physical Anthropology* 72:485–514.

White, T. D., D. C. Johanson, and W. H. Kimbel. 1981. *Australopithecus africanus:* Its phyletic position reconsidered. *South African Journal of Science* 77:445–70.

6. WHAT DO YOU STAND FOR?

Falk, D. 1990. Brain evolution in *Homo:* The "radiator" theory. *Behavioral and Brain Sciences* 13:333–81.

Hunt, K. D. 1994. The evolution of human bipedality: Ecology and functional morphology. *Journal of Human Evolution* 26:183–202.

———. 1996. The postural feeding hypothesis: An ecological model for the evolution of bipedalism. *South African Journal of Science* 92:77–90.

———. 1998. Ecological morphology of *Australopithecus afarensis.* In E. Strasser, ed., *Primate locomotion,* pp. 397–418. New York: Plenum.

Jablonski, N. G., and G. Chaplin. 1993. Origin of habitual terrestrial bipedalism in the ancestor of the Hominidae. *Journal of Human Evolution* 24:259–80.

Jolly, C. J. 1970. The seed-eaters: A new model of hominid differentiation based on a baboon analogy. *Man* 5:1–26.

Langdon, J. H. 1997. Umbrella hypotheses and parsimony in human evolution: A critique of the aquatic ape hypothesis. *Journal of Human Evolution* 33:479–94.

Leakey, R., and R. Lewin. 1978. *People of the lake: Mankind and its beginnings.* New York: Doubleday.

Lovejoy, O. W. 1981. The origin of man. *Science* 211:341–50.

Rose, M. D. 1976. Bipedal behavior of olive baboons (*Papio anubis*) and its relevance to an understanding of the evolution of human bipedalism. *American Journal of Physical Anthropology* 44:247–61.

———. 1984. Food acquisition and the evolution of positional behavior: The case of bipedalism. In D. J. Chivers, B. A. Wood, and A. Bilsborough, eds., *Food acquisition and processing in primates,* pp. 509–24. New York: Plenum.

Rose, L. M., and F. Marshall. 1996. Meat eating, hominid sociality, and home bases revisited. *Current Anthropology* 37:307–38.

Stanford, C. B., and J. S. Allen. 1991. On strategic storytelling: Current models of human behavioral evolution. *Current Anthropology* 32:58–61.

Tuttle, R. H. 1975. Parallelism, brachiation, and hominoid phylogeny. In W. P. Luckett and F. S. Szalay, eds., *The phylogeny of the primates: A multidisciplinary approach*, pp. 447–80. New York: Plenum.

———. 1981. Evolution of hominid bipedalism and prehensile capabilities. *Philosophical Transactions of the Royal Society of London* 292 (series B): 89–94.

Washburn, S. L. 1960. Tools and human evolution. *Scientific American* 203:62–75.

Wheeler, P. E. 1991. The influence of bipedalism on the energy and water budgets of early hominids. *Journal of Human Evolution* 21:117–36.

7. THE SEARCH FOR MEAT

Blumenschine, R. J. 1987. Characteristics of an early hominid scavenging niche. *Current Anthropology* 28:383–407.

Boesch, C., and H. Boesch. 1989. Hunting behavior of wild chimpanzees in the Taï National Park. *American Journal of Physical Anthropology* 78:547–73.

Bunn, H. T., and J. A. Ezzo. 1993. Hunting and scavenging by Plio-Pleistocene hominids: Nutritional constraints, archaeological patterns, and behavioural implications. *Journal of Archaeological Science* 20:365–98.

Cordain, L., et al. 2000. Plant-animal subsistence rations and macronutrient energy estimations in worldwide hunter-gatherer diets. *American Journal of Clinical Nutrition* 71:682–92.

Isaac, G. L. 1978. The food-sharing behavior of proto-human hominids. *Scientific American* 238:90–108.

Isaac, G. L., and D. C. Crader. 1981. To what extent were early hominids carnivorous? An archaeological perspective. In R.S.O. Harding and G. Teleki, eds., *Omnivorous primates*, pp. 37–103. New York: Columbia University Press.

Kaplan, H., K. Hill, J. Lancaster, and A. M. Hurtado. 2000. A theory of human life history evolution: Diet, intelligence, and longevity. *Evolutionary Anthropology* 9:156–85.

Milton, K. 1999. A hypothesis to explain the role of meat-eating in human evolution. *Evolutionary Anthropology* 8:11–21.

Shipman, P. 1986. Scavenging or hunting in early hominids. *American Anthropologist* 88:27–43.
Shipman, P., and A. Walker. 1989. The costs of becoming a predator. *Journal of Human Evolution* 18:373–92.
Stanford, C. B. 1996. The hunting ecology of wild chimpanzees: Implications for the behavioral ecology of Pliocene hominids. *American Anthropologist* 98:96–113.
———. 1999. *The hunting apes: Meat-eating and the origins of human behavior.* Princeton, N.J.: Princeton University Press.
———. 2001. A comparison of social meat-foraging by chimpanzees and human foragers. In C. B. Stanford and H. T. Bunn, eds., *Meat-eating and human evolution,* pp. 122–40. Oxford: Oxford University Press.
Tanner, N. M., and A. L. Zihlmann. 1976. Women in evolution, part 1: Innovation and selection in human origins. *Signs: Journal of Women, Culture, and Society* 1:585–608.
Washburn, S. L., and C. Lancaster. 1968. The evolution of hunting. In R. B. Lee and I. DeVore, eds., *Man the Hunter,* pp. 293–303. Chicago: Aldine.
Wrangham, R. W., et al. 1999. Cooking and human origins. *Current Anthropology* 40:567–94.

8. BETTER BIPEDS

Binford, L. R. 1981. *Bones: Ancient men and modern myths.* New York: Academic Press.
———. 1987. Were there elephant hunters at Torralba? In M. H. Nitecki and D. V. Nitecki, eds., *The evolution of human hunting,* pp. 47–105. New York: Plenum.
Blumenschine, R. J. 1987. Characteristics of an early hominid scavenging niche. *Current Anthropology* 28:383–407.
Brain, C. K. 1981. *The hunters or the hunted?* Chicago: University of Chicago Press.
Bunn, H. T., and J. A. Ezzo. 1993. Hunting and scavenging by Plio-Pleistocene hominids: Nutritional constraints, archaeological patterns, and behavioural implications. *Journal of Archaeological Science* 20:365–98.
Bunn, H. T., and E. M. Kroll. 1986. Systematic butchery by Plio/

Pleistocene hominids at Olduvai Gorge, Tanzania. *Current Anthropology* 27:431–52.

Cann, R. L., M. Stoneking, and A. C. Wilson. 1987. Mitochondrial DNA and human evolution. *Nature* 325:31–36.

Dean, C., et al. 2002. Growth processes in teeth distinguish modern humans from *Homo erectus* and early hominids. *Nature* 414:628–31.

Dominguez-Rodrigo, M., et al. 2001. Woodworking activities by early humans: A plant residue analysis on Acheulian stone tools from Peninj (Tanzania). *Journal of Human Evolution* 40:289–99.

Dorit, R. L., H. Aksashi, and W. Gilbert. 1995. Absence of polymorphism at the Zfy locus on the human Y chromosome. *Science* 268:1183–85.

Gabunia L., et al. 2001. Dmanisi and dispersal. *Evolutionary Anthropology* 10:158–70.

Krings M., et al. 1997. Neandertal DNA sequences and the origin of modern humans. *Cell* 90:19–30.

O'Connell, J. F., and K. Hawkes. 1988. Hadza hunting, butchering, and bone transport and their archaeological implications. *Journal of Anthropological Research* 44:113–61.

Ovchinnikov, I. V., et al. 2000. Molecular analysis of Neanderthal DNA from the northern Caucasus. *Nature* 404:490–3.

Potts, R. 1984. Home bases and early hominids. *American Scientist* 72:338–47.

Potts, R., and P. Shipman. 1981. Cutmarks made by stone tools on bones from Olduvai Gorge, Tanzania. *Nature* 291:577–80.

Relethford, J. H. 1995. Genetics and modern human origins. *Evolutionary Anthropology* 4:53–63.

Schick, K. D., and N. Toth. 1993. *Making silent stones speak.* New York: Simon and Schuster.

Shipman, P., and A. Walker. 1989. The costs of becoming a predator. *Journal of Human Evolution* 18:373–92.

Speth, J. D., and E. Tchernov. 2001. Neandertal hunting and meat-processing in the Near East. In C. B. Stanford and H. T. Bunn, eds., *Meat-eating and human evolution*, pp. 52–72. Oxford: Oxford University Press.

Stringer, C. B., and P. Andrews. 1988. Genetic and fossil evidence for the origin of modern humans. *Science* 239:1263–68.

Thorne, A. G., and M. H. Wolpoff. 1981. Regional continuity in

Australasian Pleistocene hominid evolution. *American Journal of Physical Anthropology* 55:337–49.

———. 1992. The multiregional evolution of humans. *Scientific American* 226:76–83.

Walker, A., and P. Shipman. 1997. *The wisdom of the bones.* New York: Vintage.

Wolpoff, M. H. 1989. Multiregional evolution: The fossil alternative to Eden. In P. Mellars and C. B. Stringer, eds., *The human revolution: Behavioural and biological perspectives on the origins of modern humans*, pp. 62–108. Princeton, N.J.: Princeton University Press.

Wolpoff, M. H. 1989. The place of Neanderthals in human evolution. In Erik Trinkaus, ed., *Biocultural Emergence of Humans in the Later Pleistocene*, pp. 97–141. Cambridge: Cambridge University Press.

9. SKY WALKERS

Dunbar, R.I.M. 1992. Neocortex size as a constraint on group size in primates. *Journal of Human Evolution* 20:469–93.

Foley, R. A., and P. C. Lee. 1989. Finite social space, evolutionary pathways, and reconstructing hominid behavior. *Science* 243:901–6.

McGrew, W. C. 1992. *Chimpanzee material culture.* Cambridge: Cambridge University Press.

Smolker, R. A., et al. 1997. Sponge carrying by dolphins (Delphinidae, *Tursiops sp.*): A foraging specialization involving tool use? *Ethology* 103:454–65.

Stanford, C. B., and J. S. Allen. 1991. On strategic storytelling: Current models of human behavioral evolution. *Current Anthropology* 32:58–61.

Tooby, J., and I. DeVore. 1987. The reconstruction of hominid behavioral evolution through strategic modeling. In W. G. Kinzey, ed., *The evolution of human behavior: Primate models*, pp. 183–238. Albany: State University of New York Press.

Zimmer, C. 1999. *At the water's edge: Fish with fingers, whales with legs, and how life came ashore but went back to sea.* New York: Touchstone.

INDEX